Cromer Fire Brigade

1881–2006

Research by Jamie Edghill
Text by Keith Entwistle

Poppyland Publishing

Copyright © 2006 Jamie Edghill, Keith Entwistle

First published in 2006
by Poppyland Publishing,
Cromer, NR27 9AN
www.poppyland.co.uk

ISBN 0 946148 81 3
978 0 946148 81 3

All rights reserved. No part of this publication may be reproduced, stored in a retrieval system or transmitted by any means, electronic, mechanical, photocopying, recording or otherwise, without the written permission of the publishers.

Designed & typeset in 11/14pt Syndor and Times New Roman by Watermark, Cromer, NR27 9HL

Printed by The Complete Product Company, Diss, Norfolk

Acknowledgements

Many thanks to the following for kind permission to use photographs and reports and for help in other ways in the production of this book:

Eastern Daily Press
North Norfolk News
Cromer Museum
Cromer Town Council
Former Chief Fire Officer Bryan Smith
Norfolk Fire Service
RNLI Henry Blogg Museum
Randall-Salter Magic Lantern Collection
Poppyland Photographs

Mr I. Alexander, Mrs E. Balls, Mr N. Babbage, Mr T. Barnes, Mr W. Cox, Mr & Mrs J. Davies, Mr G. & Mrs R. Fisher, Mr G. Johnson, Mr H. Khalil, Mrs J. Leeds, Mrs E. McHugh, Mr D. Pope, Mr T. Rogers, Mr P. Stibbons

Sponsors

Cromer Town Council
Simon Purslow Chartered Building Surveyors (Cromer)
Bond Street Antiques and Jewellery (Cromer)
Le Moon Chinese Restaurant (Cromer)
Mackinnon Construction Ltd (Cromer)
Mr & Mrs J. Leeds (Cromer)
G.S. Lane Decorators Supplier (Cromer)
Cambridge House Hotel (Cromer)
Mary Janes Fish Bar (Cromer)
The Blue Danube Café (Cromer)
Stenson Guest House (Cromer)
Virginia Court Hotel (Cromer)
Bond Street Bakers (Cromer)
Benbows (Cromer)
H.K. Discount and Clearance Depot (Cromer)
Highfield Residential Care Home (Cromer)
Hair Studio One (Cromer)
Foundations Estate Agents (Cromer)
Cromer Crab Company
Cromer and District Funeral Services
Cromer Carpets
Holland & Holland Cromer Property Corporation Ltd
Major A. R. Gurney (Northrepps)
Ivy Farm Holiday Park (Overstrand)
Overstrand Garden Centre
Fish and Chip Shop (Roughton)
Constantia Cottage Restaurant (East Runton)
The Poplars Caravan and Chalet Park Ltd (East Runton)
Wyndham Park Caravan Site (East Runton)
Shell UK (Bacton)
Mr Keith Barker
Leake's Caravan Park (East Runton)

Foreword

It gives me great pleasure to write this foreword for Jamie Edghill's book on the Cromer Fire Service.

Throughout its history Cromer has maintained an excellent record of firefighting and fire safety and from the early days the volunteers through to the now Retained Fire Service have dedicated their time to protect the people of Cromer and Norfolk. Whilst times have changed somewhat since those early days, firefighters have always required special skills to take on the risks that they face. It's not just firefighting now that has to be considered but also dealing with road accidents, chemical incidents and even terrorist situations which demand an ever-increasing knowledge.

Although there may be better information and modern technology to help deal with fires and other incidents, the inherent risks are still there. Occasionally the ultimate sacrifice will be made and a firefighter will lose his or her life.

I wish therefore to pay tribute to all the men and women who over the years have so willingly and selflessly given their time to protect us all.

Credit must also be given to Jamie Edghill, an ex-leading Firefighter at Cromer, for the splendid work he has carried out in researching the history of Cromer Fire Service and producing this excellent record for future reference.

Bryan E Smith
Chief Fire Officer for Norfolk, 1990–2000

Contents

Firefighting in Britain	4
Cromer Fire Brigade: the beginnings	6
Between the wars	22
World War 2	44
The Post-war period	60
Notes from a personal archive	67
Index	95

Firefighting in Britain
A brief outline history

The first recorded measure to prevent destruction through fire was taken by William the Conqueror in the eleventh century. He ordered the ringing of the Curfew Bell, which required townspeople to extinguish any flame – source of light and heat – at eight in the evening, to prevent an unattended fire spreading in the hours of night. Indeed, the origin of the word 'curfew' is 'couvre feu' – French for 'cover the fire'.

Following the Great Fire of London in 1666, the first Fire Insurance Company was established by Nicholas Barbon, and the first fire brigade was formed in 1680 by Barbon's Fire Office. Business was good and soon many other companies were formed, each with its own brigade. Householders had to pay an annual fee to their chosen company for the services of their fire engine in case of need; a metal plate attached to the house – a Fire Mark – demonstrated to which company the appropriate dues had been paid. The shout of 'Fire!', however, brought more than one company to the scene, and if the first to arrive discovered that the house in question was insured with a rival company, the house might well be allowed to burn down. Further, if the house were uninsured, the companies would fight each other for the rights to offer assistance – before fighting the fire, of course. It was a while before they learnt to co-operate.

In 1708, Parliament passed the Parish Pumpers Act, which required the churchwardens in each parish to purchase a fire engine and to keep it in readiness; they also had to set up blocks or taps in any piped water system available – early fire hydrants, in effect. However, these measures were implemented in a very inconsistent way.

With the fast growing urban areas, the lack of municipal direction was leading to fires becoming uncontrollable hazards. The first city to promote a more coherent response was Edinburgh, where a series of devastating fires led a certain young man, James Braidwood, to make an approach to the city authorities; Braidwood had been apprenticed to the building trade and was soon applying his knowledge of building structures to devise strategies to fight fires. Further, he managed to persuade the authorities to combine with the insurance companies to create a municipal brigade under his leadership.

Braidwood's methods were so successful that in 1833 he was asked to become the first Superintendent of The London Fire Engine Establishment, London's equivalent of the force pioneered in Scotland. For nearly 30 years he drove forward the development of firefighting services in London in terms of equipment, planning, prevention measures, training and leadership. From his early days in Edinburgh, he had understood that successful firefighting depended on some key practices: it is necessary to enter a burning building to get at the seat of a fire; firemen should never work singly, they should carry rope to effect an escape from a building and they must train at night to familiarise themselves with working in the dark. Braidwood brought practical leadership and a clear vision to his role; his work laid the foundations of the modern Fire Service.

Braidwood was directing operations at the terrible Tooley Street Warehouse fire in June 1861 when he went to support a group of his men. Suddenly, a huge explosion forced outwards the front wall of the warehouse, several storeys high, and he and a colleague both died when they failed to escape the falling masonry. He was mourned as a national hero, with a crowd one and a half miles long following his funeral cortège. Messages of sympathy were received from Queen Victoria and from all corners of the British Empire. From Australia a letter was received which said: 'On receipt of the sad news, our large fire-bell was tolled, the British ensign hoisted half-mast high, and crepe attached to the firemen's uniform, as a token of respect for one of the noblest and most self-denying men that ever lived, who spent and lost his life in the service of his fellow creatures.'

Cromer Fire Brigade

The beginnings

Cromer in 1881 was on the brink of a period of expansion which changed a destination of modest aspirations into one of the most fashionable locations in the land. The British Empire was at its height, Queen Victoria had been on the throne for over 40 years, the scope of the nation's industry was unparalleled. In Cromer, the first railway station had been opened in 1877 at the top of the Norwich road, bringing as visitors to north Norfolk large numbers of people for whom it was a new, exotic destination. The first large villas were being built on Suffield Park and the more far-sighted hotel proprietors were looking to expand and build new properties. Power, wealth and civic pride characterised the national mood.

We are very fortunate to have a range of original documents available to give us an insight into the beginnings of Cromer's Fire Brigade. We have some names and faces. We may read the first booklet produced to outline some aspects of the brigade's activities, and the framework within which everything was conducted. It was written by Edward Raven Priest, listed as Secretary and Treasurer.

A glance at the front page of Edward Priest's handbook takes us straight into that world. We will look first at the administration, at this stage under the control of a committee headed by the Vicar. It is interesting to observe that the terminology employed is not fully developed – the last three listed are simply

RULES AND REGULATIONS

COMMITTEE

The Vicar – Chairman
The Churchwardens
The Guardians
The Overseers
The Surveyors
(with power to add to their number)
The Secretary and Treasurer

THE BRIGADE

Hon. Captain
Brigade Captain
Two Foremen
Two Firemen
Four Hosemen
Three Supernumeraries

LEFT: *Cromer Fire Brigade posed with their 1886 Merryweather fire pump outside a flint barn, probably on The Mount in Cromer.*

BELOW: *An early photograph of the Cromer team in harness.*

'supernumeraries' (extras). In the side panel we may read a description of their uniform, and this earliest photo of the brigade gives us its appearance.

Priest's handbook has these sections:

- Rules and Regulations
- The Brigade (job specifications)
- General observations
- Plain Directions for the Preservation of Life in the Event of Fire
- To Bystanders
- Hydrant Drill with Hosereel
- Hydrants *(none listed, but space allocated for hand entries)*

The content of the handbook makes it clear that the Cromer Brigade is starting from scratch, as it were; much of it reads as plain common sense derived from a limited experience of firefighting. Some entries seem rather amusing 125 years later. Here comes a stern warning for any potential recruit to the brigade:

Any member guilty of practical joking or disorderly conduct at the drills or other assemblies of the brigade shall be liable to be fined by the Committee, such fine not to exceed ten shillings.

The section 'General Regulations' opens with a pleasing concern for the welfare of the general public in the vicinity of water jets.

It cannot be too strongly impressed upon the men (whether in practice or at a fire), to avoid, if possible, playing on any of the bystanders.

The Captain's job specification reads thus:

He shall attend and drill the brigade at the stated periods; and the moment the alarm of fire is raised he shall repair to the Brigade Station or spot, with all possible speed and take command of the force. When there he will be responsible for the general conduct of the brigade under his charge; he shall keep an account of the time and the number and names of the men employed, and will not suffer idlers to stand about and interfere with his men.

UNIFORM

CAPTAINS – Helmet with distinguishing mark, tunic, trousers, Wellington or Napoleon boots; may wear gorgette, epaulettes, whistle and chain.

FOREMEN – Helmet, leather with brass mountings and brass comb; tunic with one epaulette, trousers etc, leather belt with axe and case and hand-lamp.

FIREMEN – Helmet, leather with brass mountings, tunic, trousers etc, leather belt with hand-axe in case, lifeline, hand-lamp.

HOSEMEN – Helmet, leather with brass mountings; tunic, trousers etc, leather belt, wrench, hand-line.

RESERVE-MEN – Undress Fireman's cap and belt.

There is keen awareness of the welfare of horses:

When a stable is on fire, the horses should be immediately removed; the usual harness put on, the eyes bandaged, a damp or wet cloth placed over the nose or mouth, then led out as if they were going to their regular work; and they will generally do so without difficulty. This has been done on several occasions with great success.

Early years

Early histories show a slight confusion as to the actual year of founding, with Savin giving it as 1880, while Brigade Captain Frost, in a summary written in his Annual Report of 1899, gives it as 1881.

Gentlemen,

In presenting this report, which is made up to the end of 1899, it may be of interest to note that the Cromer Fire Brigade was established in the year 1881, and was at first under the management of a Committee appointed by the Vestry, and supported by voluntary subscriptions assisted by sums voted from the Poor Rate. The control of the Brigade passed into the hands of the Local Board on its formation in the year 1884. The Brigade was formed for hydrant service only and it was not until 1886 that the present steamer was purchased. The cost was principally defrayed by subscriptions, the local Board making up the deficiency of about £50. The Brigade was first housed at premises in the 'Mount', but has since 1891 occupied the present station, which is rented by the District Council from the Cromer Town Hall Company. The Fire Escape was purchased by the District Council in 1896.

RIGHT: *The Merryweather steam pump with attendant somewhere in the Cromer area. Steam pressure was used to force the water through the hoses.*

CROMER URBAN DISTRICT COUNCIL.

FIRE BRIGADE.

NOTICE!

STRONG ACTIVE YOUNG MEN required for the Brigade, which is being increased in numbers, and to fill up vacancies.

Particulars can be obtained at the office of the Council, and personal application to be made at the Brigade Station on Friday, Mar. 26th, at 8 p.m.

JAMES K. FROST,

Cromer, *Captain.*
March, 1897.

The original specification for the steamer describes it as 'similar to those in use and approved by the Fire Brigades of Staines, Windsor, Richmond, Farnborough, Appleby, Littleport, Bourne, Enfield, etc. . . .' No self-respecting town dare refuse!

> Specification of Merryweather's Patent Horizontal Steam Fire Engine.
>
> An Engine similar to those in use and approved by the Fire Brigades of Staines, Windsor, Richmond, Farnborough, Appleby, Littleport, Bourne, Enfield &c. Of the patent long stroke direct acting pattern having few working parts and those so arranged as to be readily accessible for oiling & cleaning and balanced & adjusted to avoid undue friction. The Pump to be of one solid casting in best gunmetal fitted horizontally with back and front covers and internal long and wide opening valves hinged horizontally without springs or gratings. Two delivery outlets for hose connection in front of and below Pump, so as to empty pump after working, and suction inlet at the side but near the deliveries. The Steam Cylinder to be fixed horizontally behind the Pump and fitted with slide jacket having circular slides and all the requisite pet cocks, steam valves and pipes. One turned rod to form both water and steam pistons with necessary piston blocks and rings and cupped leathers. The water end of piston cased in brass and a cross head connected to piston rod with arm

ABOVE: *The garage-like extension at the rear of the Town Hall (now Plumb Center) served to house the fire engine from 1891; subsequently an adjacent warehouse was used.*

BELOW: *The fire station built in 1905, capable of housing two appliances and with room for training in the yard.*

The new Fire Station, 1905

Later on in his report, Captain Frost urges the Committee to give serious thought to the matter of a new Fire Station.

I would again draw attention to the urgent necessity for the provision of a new Station sufficiently large to properly house all the appliances, and to provide space for drilling purposes. The existing Station is totally inadequate, and the adjacent open spaces which have hitherto been used for drilling are rapidly becoming unavailable.

It is clear from the records available that the steps to the full establishment of the Fire Brigade were taken over a number of years as funds were allocated. As regards location, we have hints of three early sites. The first comes in Savin's survey of the Brook Street area of town in the 1890s. Talking of a range of old farm buildings, he says,

This row of barns formed one side of the driftway mentioned before. An old Fire Engine, with hand lever pumps, was kept there; it belonged to Mr Sandford.

(Henry Sandford, a leading citizen of Cromer, was a coal merchant who had a yard at the top of the Gangway.)

And again,

Returning to Crisp's old house, a large gateway led to the back of the house and various buildings which bordered the road up to Flint House; William Murrell had Livery Stables here, he used to drive one of the coaches between Cromer and Norwich. When the Cromer Fire Brigade was started they used one of the old coach houses to store their equipment.

The premises in the Mount mentioned in Captain Frost's report of 1899 would later disappear as Mount Street was built up.

The third, occupied at the time Captain Frost was writing in 1899, was the rear of Cromer Town Hall.

By 1901 they had already earmarked a site opposite them in Canada Road, just a little higher up,

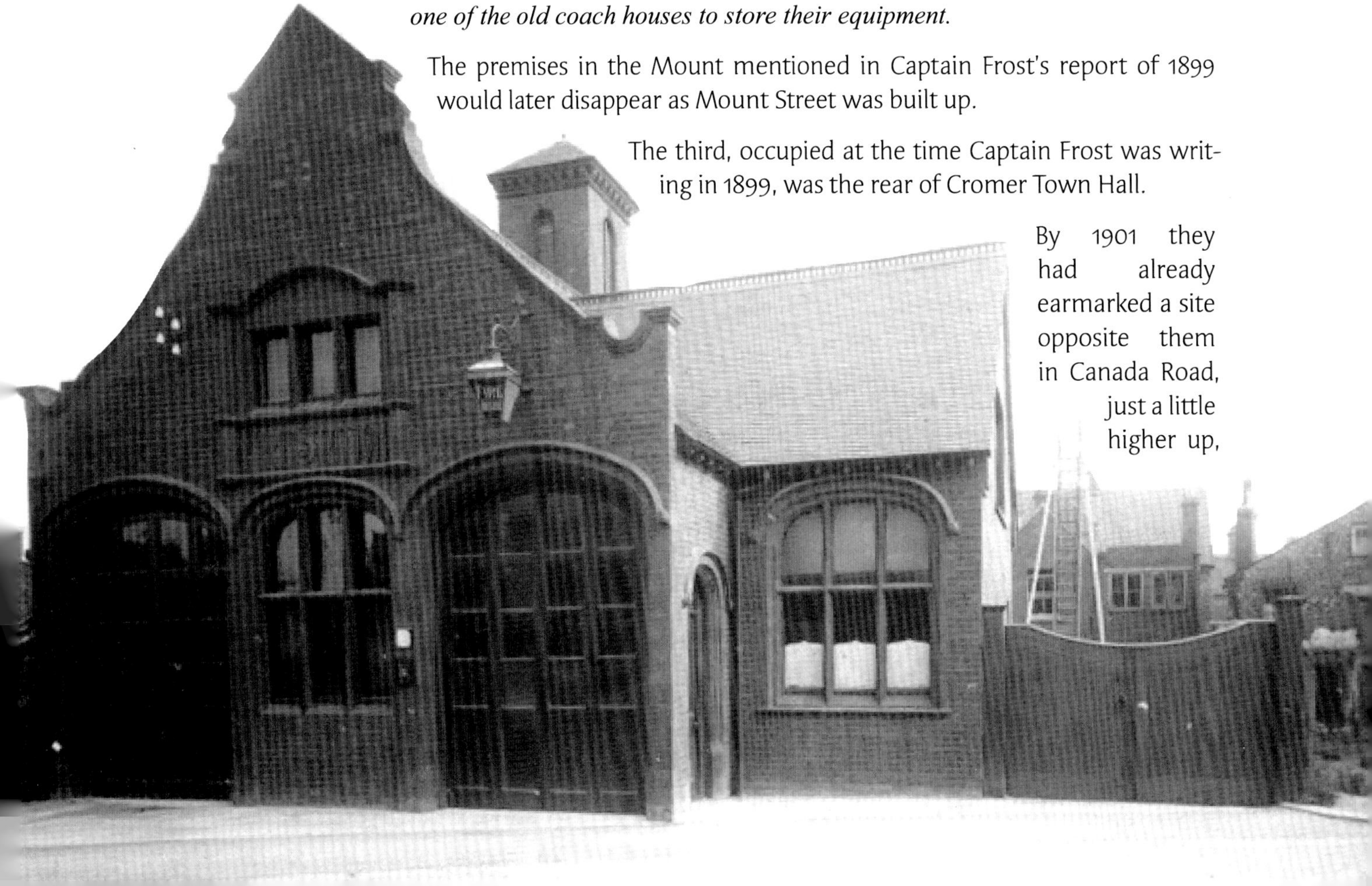

This poster announces an inquiry prior to the development of two sites: the Council offices (which were to occupy the building now Lloyds TSB) and the new fire station on Canada Road (now the Kingdom Hall of Jehovah's Witnesses).

as being ideal for their purposes.

Two more years passed before the next step in the saga; the brigade was forced to leave its existing rented premises and move into an adjoining warehouse property. This probably forced the issue, and plans were drawn up in early 1905 for the new station to be built. The initial plan included the following:

Proposed Fire Station

The site has a frontage of 54 feet to Canada Road. The appliances possessed by the Brigade include steam fire engine, hose reel and cart, and fire escape. The existing building is not large enough to house all the existing appliances, the escape having to stand outside. It is proposed to convert the escape, which is at present propelled by hand, into a horsed escape and it is intended to do this by the adoption of Messrs Merryweather's dogcart escape. This will enable the escape to be housed in the new station and the dogcart can be used in connection with the escape or independently as a hosereel cart for the outlying parts of the town.

The new premises will comprise:

¶ *Engine room, with small watch-room, tower and workshop, and hose-drying tower.*

¶ *On the first floor accommodation will be provided for a caretaker who will probably be a member of the Brigade.*

¶ *There will be a yard at the rear which will be available for drilling purposes. At present the Brigade have no ground for this purpose.*

Some early incidents

We have had access to the incident book listing Cromer Brigade's attendance at fires in these early years. In most cases, sadly, very little detail is available, but sometimes other sources reveal some interesting aspects. For example, the newspaper article celebrating 50 years' service of Jack Kirby, in 1935, includes the following report.

THE PUBLIC HEALTH ACT, 1875.

CROMER

WHEREAS the Urban District Council of Cromer have applied to the Local Government Board for sanction to borrow sums of £3,500 and £500 for purposes of the proposed Council Offices at the junction of West Street and Chapel Street and for the widening of West Street, respectively, and £1,500 for the purchase of land at Canada Road and the erection thereon of a Fire Brigade Station; and the Local Government Board have directed Inquiry into the subject-matter of such Applications:

NOTICE IS HEREBY GIVEN that M. K. North, Esquire, M.Inst.C.E., the Inspector appointed to hold the said Inquiry, will attend for that purpose at the Council Offices, Town Hall, Cromer, on Saturday, the Sixteenth day of January, 1904, at a Quarter to Eleven o'clock in the Forenoon, and will then and there be prepared to receive the evidence of any persons interested in the matters of the said Inquiry.

Local Government Board,
6th January, 1904.

S. B. PROVIS,
Secretary.

Mr Mace, one of the Cromer ratepayers and a photographer in the town, was amongst those who took exception to the quality of the buildings, with accompanying cost, undertaken by the Urban District Council at the beginning of the twentieth century. He produced a booklet with his comments and photographs of the buildings. It had evidently been intended that new council offices, with a fire station attached, should have been built, but concerns about the cost led to the Fire Station becoming a separate project – though he still considered it over-elaborate.

Beckham 'Palace' (workhouse!) in its heyday.

Beckham Palace 1ST APRIL 1888

The Cromer Brigade's first big fire in Mr Kirby's young days was at Beckham Workhouse on Sunday, 1st April 1888. The Cromer firemen were roused by the ringing of the church bells, the custom in those days. But one fireman who lived near where the brigade assembled remembered that it was All Fools' Day. 'You aren't going to make a fool of me,' he shouted from his window.

Major Barclay's stables 16TH DECEMBER 1899

A handwritten report, dated 12th January 1900, from the Brigade Captain describes a fire at the premises of a near neighbour of his at The Grange in Cliff Avenue.

Gentlemen,

I beg to report that on the 16th December the Brigade was called to a fire at Major Barclay's stables, The Grange, Cliff Avenue. The alarm was given at my house by a servant from the Grange at about 7.30 p.m. and I at once despatched a messenger to call out the Brigade. I proceeded direct to the scene of the fire and found that the fire was burning in the loft over the stables. The steamer arrived quickly but a stand pipe not having been placed on the steamer a messenger had to return to the station for it so that until his return no water was available. The roof over the stables was destroyed and also over part of the harness room, but otherwise the harness room, coach house and stables were saved.

Stables in Hanworth 11TH OCTOBER 1900

Here is the entry from the Incident Book:

Called to fire at Hanworth at Mr Richard Chapman's builders' premises. Alarm given by messenger on horseback at 10.30 p.m. Brigade summoned by electric calls, captain by messenger. Engine arrived Hanworth, Captain, 1st Lieutenant and Assistant engineer Warner followed. Supply of water obtained from pond on common. 1 line hose used and afterwards 2 branches. Fire confined to building in which same originated, a building of 2 stories [sic] used as carpenters' and carvers' shops. Brigade remained by fire till 3 a.m., 2 men remaining on duty till 8.30 a.m. Origin of fire unknown.

It is quite interesting to compare the above notes with the following words from Captain Frost's report on the incident in his annual summary of the brigade's activities:

The alarm was given at 10.30 p.m, and the Brigade, which was summoned by Electric Calls, turned out very smartly, reaching Hanworth shortly after 11 p.m.

This seems a remarkable claim, given that all of the following had to be ac-

This splendid impression of Cromer Fire Brigade responding to a call is thought to represent this incident at Hanworth. It appeared in the Illustrated London News.

A modern photograph of Chapman's premises in Hanworth

complished in that brief half-hour. The message received, the coachman had to be alerted; he then had to harness his horses, which were kept on Beef Meadow opposite Cromer Hall, lead them to the brigade headquarters, put them into the shafts, then head full tilt for Hanworth. This involved a steady climb out of Cromer, and a total journey of over six miles. A slight exaggeration, Captain Frost?

Sheringham Lodge 28TH JANUARY 1905
4.30 p.m. Called to fire at Sheringham Lodge by messenger from Mr Warby's. Steamer horsed by Palmer. Captain went ahead in Mr Bullen's trap. Water supply intermittent by carts. Building practically destroyed on arrival but burning parts knocked out to woodwork and dangerous gable pulled down.

Roughton Mill 17TH SEPTEMBER 1906
Steamer used. 4 hours away. Supposed causes – friction from sails breaking loose in gale. Interior gutted and connecting buildings. (Roughton not within area and brigade attended in error)

Claim for expense recovered £11 10s 6d

Hanworth Hall 2ND SEPTEMBER 1909
Called to fire at Hanworth Hall by telegram from Colonel Barclay. Chairman of Committee authorized sending engine as negotiations pending.

3 p.m. – telegram despatched Hanworth

3.17 p.m. – received Cromer

3.22 p.m. – delivered office

3.35 p.m. – engine left with engineer; Chief Officer and Lieutenants motored over and men followed.

Roughton Mill

Hose insufficient and Colonel Barclay despatched motor for more. In meantime hose connected to owner's small hose and supply from pit on common pumped on building. On additional hose arriving it was coupled up and 2 jets used. Buildings consisted of dynamo house and saw engine shed etc burnt out. Adjoining buildings and stock in yard saved.

Expenses £16 16s 6d paid by Norwich Union.

Cromer Hall Farm 30TH NOVEMBER 1909
4.55 p.m. Chimney fire.

Red Lion Hotel 30TH NOVEMBER 1909
5.45 p.m. Call to fire at Red Lion Hotel caused by gas explosion.

Neighbouring parishes

Cromer itself was well set with its Fire Brigade, financed primarily out of the municipal purse. But the issue of its availability for use by communities outside the town boundaries raised the question: who foots the bill? From very early on, there were attempts to levy an annual charge from neighbouring parishes, but it was always on a voluntary basis and was to be the cause of considerable friction over many decades, as we shall see. The minute-books give ample evidence of the fluctuating allegiances of the parishes. The matter was first raised by Cromer's Urban Clerk in letters to Sir Samuel Hoare and J. H. Gurney, both of whom were personal subscribers to the Fire Brigade.

Here is the text of the Clerk's letter:

Dear Sir,

The Cromer Fire Brigade have hitherto held themselves open to attend all fires in the adjacent district, but the Fire Brigade Committee have recently been considering whether this practice should not, to some extent, be modified.

The expense of the upkeep of the brigade (amounting to a considerable sum each year) is borne by the Cromer ratepayers and the committee thinks that if the brigade attends fires in parishes outside the Urban District those parishes should pay an annual sum towards the maintenance expenses under the provisions of the Parish Fire Engine Act, 1898.

I believe the Norwich Town Council have made an agreement with several of the parishes surrounding the city under this Act. A difficulty has however arisen here, owing to the fact that when subscriptions were invited for the Steam Fire Engine it was stated that donors of £10 and upwards would be

---

From minute-book, 4th February 1913:

Motion passed to the effect that 'Church bells to be rung in all cases of alarms of fire'. Moved by Mr Savin.

entitled to the use of this engine free of charge, and before communicating with any parish authorities the committee have instructed me to obtain the opinion of any donors having property within such parishes.

I find on referring to the list you gave a subscription of £20 and I should be glad if you would inform me whether you see any objection to the parishes of, say, Sidestrand, Northrepps, Southrepps, Beeston Regis being asked to contribute in the manner suggested.

Yours faithfully,

Clerk

Sir Samual Hoare's reply underlines his firm commitment to the idea of Fire Protection.

Fire Engine

When I subscribed to the above I did so on the understanding that it would be available in case of fire on my property in the neighbourhood – but I shall not allow this to interfere in any arrangement you may be able to make with the Parishes in which I am personally interested. I originally promoted the Bill . . . under which parishes are able to unite (?) for protection for fire, and I am strongly in favour of its being put in force.

Emboldened by the positive response of Sir Samuel Hoare, the Clerk wrote next to J. H. Gurney, Esquire, of Keswick Hall, Norwich, owner of Northrepps Hall. The early part of his letter is identical to that sent to Sir Samuel, but he adds a further paragraph.

I may say that I have heard from Sir Samuel Hoare, Bart, MP, who was also a subscriber, that he should not allow the fact of his having subscribed to the engine to interfere in any arrangements the Council might be willing to make with the parishes in which he owns property.

However, Mr Gurney saw things rather differently!

Dear Mr Frost,

The £20 subscription to the fire engine was given by me, but I really do not think the engine would be of much use to our property at Northrepps. Northrepps Hall has got an engine of its own, and the other houses have got no ponds and very little water in the wells of late years. The village pond on the

> KESWICK HALL,
> NORWICH.
>
> 21st April 190[?]
>
> Dear Mr Frost
>
> The £20 subscription to the fire engine was given by me, but I really do not think the engine would be of much use to our property at Northrepps. Northrepps Hall has...

Norwich Road is nearly always dry, and if the houses catch fire they will have to burn. For all practical purposes Northrepps may be excluded from the area which the engine will have to serve in case of fire, but it is quite fair that parishes with a water supply should subscribe to its maintenance.

Chief Officer's reports

Reading between the lines in the annual reports written by Brigade Captain J. K. Frost is a fascinating exercise. Each year he provided a summary of the year's activities, including the incidents attended, the provision of new equipment and progress reports on matters under development (the new Fire Station, for example). The report from 1900 includes mention of three key aspects of the Brigade. It identifies new Alarm Posts (Fire Alarms) created in the town.

During the year the Council decided to provide Fire Alarm Posts under rental from the General Post Office at Suffield Park (Park Road), Runton Road and Norwich Road, and these have since been fixed.

It includes a stocktaking.

The appliances now possessed by the Brigade include:

¶ *Steam fire engine, Merryweather's No. 1 Volunteer pattern, specified to pump 300 gallons per minute.*
¶ *Fire escape, 50 ft., with supplementary ladder to reach 60 ft.*
¶ *Hose reel.*
¶ *Hose and ladder cart.*
¶ *Scaling ladders.*
¶ *670 ft leather hose and 550 ft canvas hose.*

The cost of the above and other appliances, uniforms and accoutrements was at the last stocktaking £890 3s 6d.

And it provides statistics on the number of fires attended.

Gentlemen, during the year 1900, no calls were received in the Urban District, the only call in the year being to the builder's premises at Hanworth belonging to Mr Richard Chapman.

(See page 12.)

That hardly seems credible, but a look at figures for subsequent years does not do a great deal to support the view that the Fire Brigade was providing value for money. In 1901 there were two fires, one in Cromer and one in

The situation as regards water levels in the Northrepps pond was very different a few years later; Captain Frost's report for 1909 says that 'the services of the engine were also utilised in December for pumping out the pond on the Northrepps Road, which was overflowing on the roadway'!

Demonstration

As the poster makes clear, this demonstration took place on Thursday, 18th November 1909 – early closing day. It proved a suitable occasion, as well, to present Long Service Medals to the following members.

Hon. Captain J. K. Frost
– 28 years – Silver medal and bar
First Lieutenant A. H. Fox
– 12 years – Bronze medal
Engineer F. W. Love
– 12 years – Bronze medal
1st Assistant Engineer – F. W. Leggett
– 15 years – Bronze medal
Fireman No. 1 – J. B. Kirby
– 24 years – Silver medal
Fireman No. 2 – J. Kettle
– 26 years – Silver medal and bar
Fireman No. 3 – H. Payne
– 28 years – Silver medal and bar

The programme for this Demonstration has been preserved:

At the Station

1. Inspection of station and appliances

2. Presentation of N.F.B.U. Long Service Medals by Mrs B. B. Bond Cabbell

3. Picking up of insensible people

4. Fire Escape

    a) Use of escape (4 positions)

    b) Rescue by escape from tower

    c) Rescue of persons by lines

5. First Aid to injured

On the Meadow

6. Hydrant Drill Competition   2 teams

    First Prize 15/-   Second Prize 10/-

7. Steamer Drill Competition   2 teams

    First prize £1   Second Prize 12/6

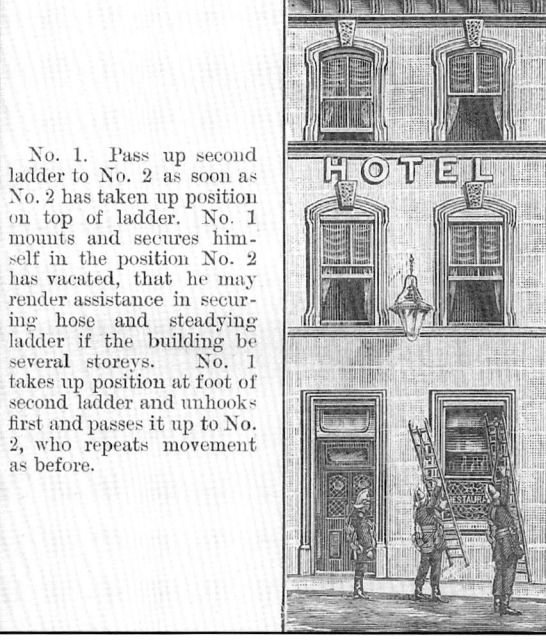
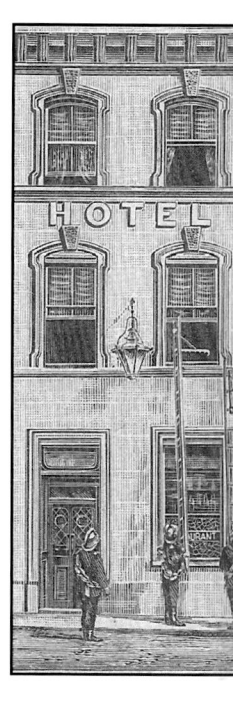

ABOVE: *Pompier ladder drill for three men, from the 1902 Drill Book of the National Fire Brigades Union.*

Sheringham which proved a false alarm; in 1902 the brigade received, again, just two calls, but in neither case were its services required; in 1903 there were three, but in 1904 none at all. In 1905, there was one from Sheringham Hall Lodge; and so it goes on. In 1909 there were all of three fires – two of which were on the same day! In contemporary terms, this really does appear to be a serious shortage of incidents requiring the services of such a handsome and expensive force, especially considering the extent of investment by the town. Perhaps a spot of public relations was called for! The Fire Brigade needed to be seen in action.

Whatever the reason, Cromer Fire Brigade went public on 18th November 1909 with an invitation to the townspeople to view the Fire Station and watch a display on the Meadow (see previous page).

This is how the Chief Officer described the year ending 31st December 1913, the last year of service before the onset of the First World War.

To the Chairman and members of the Fire Brigade Committee

Gentlemen

I have to report that during the year the Brigade received two calls, both within the Urban District. The first, to the furniture business of Messrs A. H. Fox Ltd, Prince of Wales Road, was of serious character, considerable damage being done to the stock and to the interior of the building. The brigade were able to save the structure and to prevent the spreading of the fire to the adjoining buildings.

The second call was to Mr F. W. Caley's house, Northrepps Road, which had been struck by lightning, setting the roof on fire. The brigade succeeded in getting the fire under control quickly and were able to confine it to the upper part of the building. In both cases the charges and expenses were paid by the

No. 2. Raises first ladder to first window, and after placing in position he mounts and secures himself by the hook on pompier belt and receives the second ladder from No. 1, which he fixes to the second window as before. He now mounts and proceeds to rescue, or if water is to be used he will lower his life line and pull up hose with branch attached.

No. 3. No. 3 is responsible for the working of the life lines and sending all gear aloft, and keeping the lines clear of building, so that anyone descending will not foul any projections, but on no account must he grasp the line so as to prevent it turning in his hand, or pull it taut, or it will prevent the man descending.

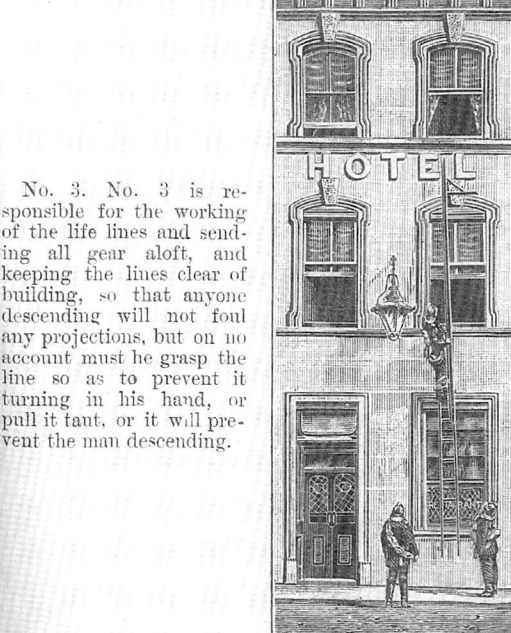

Secure the life line to some suitable fixture, then throw it out of the window, and see that it is clear, take hook and line in left hand, and with right hand take two turns round the hook with the top portion of rope. Gather in all slack and step out of the window, carrying the left hand with line well down on buttocks, and let line run through hand.

Insurance Company concerned, and in the latter the owner and tenant sent letters appreciating the work of the brigade.

All the brigade hose, stand-pipe and other fittings have now been converted into instantaneous couplings, and with this exception there has been no alteration in the appliances possessed by the brigade. Special connections have been provided by the Grand Hotel, Cromer, and the Overstrand Hotel, to enable their hose to be used with that of the brigade.

In accordance with the decision of the Council, arrangements have been made by which the church bells will be rung in case of alarms of fire.

Your committee have considered the question of the disadvantage under which the brigade laboured, especially during the summer, by reason of some of the members being absent from the town and have made arrangements for the formation of a reserve of six men who can be called upon if necessary; they will attend a quarterly drill and be paid a retaining fee of three shillings a quarter.

During the year, Mr R. W. Palmer, who has acted as coachman for the past 25 years, resigned his post and Mr R. Crown was appointed in his place. I should like to place on record my appreciation of the way Mr Palmer carried out his duties.

The conduct of the brigade has been satisfactory and I have pleasure in acknowledging the assistance at all times rendered by the Lieutenants and the Fire Alarm Superintendent.

I am, Gentlemen,
Your obedient servant
J. K. FROST
Captain

The following letters received in connection with the fire at Mr R. Caley's house were read out and ordered to be entered on the minutes.

Dear Sir,

Will you allow me to express my very warmest thanks to you and to all the members of the Cromer Fire Brigade for the very prompt and efficient manner in which you dealt with the fire at this house yesterday. I heartily appreciate all your efforts which so greatly minimised the extent of the damage.

> *With renewed thanks,*
> *Believe me, Yours very faithfully,*
> *Frederick W. Caley.*

Dear Mr Frost

Allow me to take this opportunity to thank you and every member of the Cromer Fire Brigade for the splendid service you rendered me and Mr Caley on Tuesday on speedily subduing the fire at Hilburgh. When I looked over the house on Wednesday morning and saw the serious nature of the fire I was agreeably surprised to find that the fire was mainly confined to the roof and although I was absent I quite saw that your brigade must have used every effort in their power to get the fire under control so quickly. Please accept my best thanks.

> *Yours truly*
> *S. Girling*

The Chief Officer further reported that Mr Caley had in addition sent him a cheque for three guineas towards the private funds of the brigade.

Horses

It was never intended that the Canada Road premises should include stabling in its accommodation, so any call to an incident would begin by collecting the horses. The only entry to cover this aspect of Cromer Fire Brigade in the first years of the twentieth century is to be found in a handwritten text of one edition of the Handbook.

The Brigade has harness for two horses but in nearly all cases three horses are required and for a long journey four would be necessary. The cost of providing additional harness for Leaders would be £12 5s 0d. Mr R. W. Palmer is willing to contract for supplying horses and to reserve a special place on his premises for the harness (4 horses) with 'Cromer Fire Brigade' painted over it; and would also keep the harness clean and in good condition for a

An advertisement for Mr Palmer's stables in Brook Street, from a brochure of about 1900.

payment of £2 2s 0d a year. Mr W. R. Palmer would act as driver on being paid a retaining fee of £1 0s 0d a year. I recommend that additional harness be provided and the above arrangements made with Messrs Palmer.

World War 1

Records covering this period are rather thin. The Brigade struggled to maintain its quota of manpower, so many men being away from the town on active service. In February 1915, it was decided to post a member of the Brigade at the Fire Station on fire-watching duties between 6 p.m. and 10 p.m. In October a coordinated approach was discussed with the Suffolk Regiment, currently stationed in the town. As regards the possible threat from enemy aircraft, a time-honoured solution was adopted:

The question of what precautions could be taken towards minimising danger of fires caused by aircraft was discussed, and the Chairman of the Committee and of the Council were appointed to a Sub-Committee to consider and deal with the matter.

On several occasions, attempts to hold Committee Meetings were thwarted by there not being a quorum present.

Between the wars

Reading the minute-books of Cromer Fire Brigade after World War 1 gives a picture of depletion and exhaustion – similar, no doubt, to the state of the nation at that time. We have seen how difficult it was to get enough members to attend the Committee meetings in the last months of 1918. One of the main stories of the following three years concerns trousers and boots.

10th June 1919 *The question of new uniforms was discussed and in view of the current heavy cost it was decided to make enquiries as to the possibility of securing second-hand uniforms and boots.*

9th September 1919 *The Chief Officer reported that he had been unable to obtain any second-hand uniforms and was instructed to make enquiries from Government Surplus Department and also with respect to hose.*

2nd December 1919 *The Chief Officer reported he had been unable to obtain any offers of second-hand uniforms and was instructed to endeavour to purchase cloth for making up uniforms and boots.*

2nd March 1920 *The Chief Officer submitted correspondence with Government Departments (Surplus) as to uniform materials and boots, none of which were at present available but was instructed to make further enquiries from dealers.*

1st June 1920 *The Chief Officer reported that he had not been able to obtain any second-hand uniforms but had obtained 2 pairs of new khaki trousers which were produced and which could be dyed. He had not been able to obtain any boots. The matter was directed to stand over for the present.*

7th December 1920 *The Chief Officer reported that he had been unable to obtain satisfactory second-hand boots or uniforms and submitted an estimate for new boots.*

But can we detect a hint of relaxation in the air?

7th December 1920 *It was resolved that seven pairs of new boots be ob-*

tained and that quotations be obtained for the same; the question of uniforms to further stand over temporarily having regard to the probable fall in prices.

8th March 1921 *The Clerk submitted prices of boots as under:*

Hobson and Co, London – 52 shillings, 6 pence per pair

Loads and Sons, North Walsham – 75 shillings per pair

It was resolved that seven or eight pairs of boots from Messrs Hobson & Co be ordered.

6th December 1921 *It was resolved that new uniforms be obtained for the members of the brigade at an estimated cost not exceeding £50, and that the estimate of Doan & Co Ltd for supplying tunics and trousers at 47 shillings 6 pence and 23 shillings be provisionally accepted.*

5th December 1922 *The Clerk reported the brigade had been supplied with new uniforms as authorised by the Council.*

It took over three years to finance and supply basic items of clothing for the men!

The beginning of the end for horses

The First World War did much to stimulate the development of the species 'lorry', and its potential to a Fire Brigade was soon appreciated. Here is its first mention in the minutes.

7th March 1922 *It was resolved that experiments be made with a view of ascertaining the practicability of making use of motor power through a local garage for hauling the steamer in case of fire.*

By the time of the next meeting, everything was in place.

5th December 1922 *The Clerk reported the arrangements which had been made with Messrs Rounce and Wortley for the haulage of the steamer by means of attachment to one of their motor lorries.*

Towing the steamer by somebody else's lorry, however, was only an interim solution, and it was actually the unreliability of this system that precipitated the purchase of a fire engine proper. The problem arose at a call-out to a fire at Golden Farm in Northrepps. The incident book reads:

30th June 1928 *Call received at 6.25 a.m. from Mr Golden. Captain and Lieutenant Harrison left by car and reached fire at 6.35 a.m. Carpenters shop and buildings burnt out and being got under control by buckets and water carts. Difficulty in obtaining lorry and starting same. Engine started at 7.20* *a.m., reached Northrepps Hill but sent back. Mr Golden decided the engine was not necessary at 7.15 a.m.*

There was evidently huge embarrassment all round. They had only recently (on 4th June) discussed the notion of purchasing a motor fire engine. Now they had to act; a special meeting of the Fire Brigade Committee was called three days later. A resolution was passed:

To recommend that the Council purchase a Motor Fire Engine with the least possible delay.

A fact-finding group was despatched to the works at Guildford of Dennis Bros Ltd, to view a second-hand machine at £150. But their eyes were obviously taken with a brand new 'self-propelling motor fire engine' they saw that day, because when the decision came to choose from a shortlist of tenders, it was the Dennis, costing £747 14s, which was chosen. But the matter did not end there; at their next meeting on September 3rd the Clerk read out a complaint.

A letter was received from Northrepps Parish Council asking for an explanation of the failure of the Cromer Fire Brigade in answering a call at Northrepps on 30th June and for a copy of the agreement made for the services of the Brigade.

They were clearly not at all happy with the existing state of affairs.

We have already quoted from the minute-books on various occasions, and those for this period reveal steady progress in all aspects of Fire Brigade performance. So far, we have seen this in respect of uniforms and the method of propulsion for the steamer. Over the years, there develops another issue which becomes critical as public awareness of the value of the Fire Brigades strengthens – relationships with the brigades of neighbouring towns. Until the First World War, the Cromer Fire Brigade had been the property of Cromer Urban District Council, available for use in locations beyond town boundaries on certain conditions, and through payment of certain sums. During the inter-war period we see the slow steps through which the municipal brigades start communicating with each other with a view to working together if required.

5th December 1922 *The Clerk reported that arrangements had been made with the Sheringham Fire Brigade by which the hose of the two brigades would be available for joint use if required and that the necessary special couplings had been provided. He further reported that he had not been able to effect similar arrangements with the North Walsham and Aylsham Brigades.*

Even where both parties were happy to pool resources, however, problems sometimes arose. On one occasion in 1925, the Cromer Fire Brigade responded to a request from the Holt Brigade for assistance with a blaze in the town. This is the report in the Incident Book

15h–16th November 1925 *Holt – Mr Lewis' cabinet and joinery works. Call received at 11.45 p.m. from Chairman of Holt RDC per police. Arrived at Holt at 12.50 a.m. At first not required but fire having broken out afresh*

Cromer Fire Brigade photographed outside the Fire Station in 1925
Reserves: C. J. Swann, R. J. Balls, J. Blythe
Firemen: S. Nockles, P. Hastings, W. Mayes, C. Randell, Engineer F. Love
T. Allen, A. Kirby, W. J. Lusher, W. Hawes, 2nd Assistant Engineer J. Danaher, 1st Assistant Engineer C. Swann
Alderman D. Davison, JP, OBE, Chairman of the Urban District Council
2nd Lieut. E. Harrison, 1st Lieut. R. V. Bullen, Captain T. L. Randall, Station Superintendent J. B. Kirby

> Date... Nov. 15 – 16. 1925
> Place... Holt. Mr Lewis' Cabinet & Joinery works.
> Hour—Start... 11-45 PM Dismissal... 6' A.M.
>
> Call received at 11-45 PM. from Chairman of Holt R.D.C. per Police. Arrived at Holt 12-30 A.M. At first not required but fire having broken out afresh

started pumping at 1.45 a.m. and continued intermittently until 4.45 a.m. when fire was extinguished and place left safe. Arrived back and made up 6 a.m. 1,100 ft of hose damaged by fire. Refreshments paid by captain £1 5s 0d. Taxi from garage to fetch Engineer Love (a/c to be sent to Holt RDC as early as possible).

At first sight, a simple matter of mutual assistance. But there can always be a twist, an alternative interpretation – or even a way of not having to pay up for services rendered? The Holt Fire Brigade saw things a little differently, and communicated its views in writing. A fortnight later, Cromer Fire Brigade Committee had a meeting, with hot correspondence on its agenda.

30th November 1925 *A letter from the Captain of the Holt Fire Brigade was read stating that the fire at Holt started afresh after the Cromer Brigade had left. The Clerk was instructed to reply that the fire had been extinguished by the Cromer Brigade to the satisfaction of the Captain of the Holt Brigade and left on his instructions.*

Unfortunately the means through which this little spat was resolved are not reported!

Accommodation in the Fire Station

A further step towards consolidating the place of the Brigade within the town's services was the move to convert the upper floor into living accommodation for brigade personnel. The suggestion was first made in 1921.

6th December 1921 *The Clerk reported that the Surveyor had with himself inspected the Fire Station and considered that accommodation for a resident fireman could be provided under existing roofs with a new outside staircase. It was resolved that the Surveyor be instructed to prepare a plan and estimate for carrying out the alteration.*

3rd May 1923 *The Committee considered the conditions of the occupancy of the rooms at the Fire Station. It was resolved that the rooms be offered to Mr J. B. Kirby the Fire Alarm Superintendent during the pleasure of the Council at a rent of £10 per annum inclusive of rates and charges for water, but exclusive of electric light. It was further resolved that the salary of Mr J. B. Kirby as Station Superintendent be increased from £5 to £15 per annum. Mr Kirby attended and accepted the above terms and undertook to make arrangements by which either his wife or himself would be on duty at the Station in the evenings and as far as practicable in the daytime.*

So far, so good. But over the next 12 years we may detect a touch of resentment coming into the situation. Perhaps 'living over the shop' for a net sum of £5 per year did not prove a very congenial deal for the Kirbys. The minutes read:

4th March 1935 *The question of the salary paid to the Fire Station Superintendent J. B. Kirby was raised. It was reported that he at present received £15 a year and paid £10 a year for his rooms at the Station. Before living at the Station he had received £5 a year as Superintendent and calls were received by Mr Crisp at his house adjoining the Station. By living at the station either Kirby or his wife were constantly on duty to receive calls that might be made.*

It was resolved that the Superintendent be charged no rent for the rooms at the Station.

Captain Kirby: 50 years of firefighting

The service provided by a Fire Brigade depends first and foremost on the quality of its personnel. Read the lifetime's commitment of J. B. Kirby from this newspaper report (14th January 1935), which tells of his first 50 years in uniform:

Record of Superintendent Kirby of Cromer

Fifty years a fireman. That is the proud record of Station Superintendent John Bentley Kirby of Cromer Fire brigade who has served since he was 21 years old. Now at 71 he has no thoughts of retiring. He continues his nightly duty task in case of an outbreak of fire with the same conscientiousness with which he fought fires in his early days, and when he is not there, Mrs Kirby does the duty voluntarily.

Mr Kirby is one of three men now living in Cromer who were associated with Cromer's first brigade, which was started in 1881, although Mr Kirby did not join until January 5th, 1885. He was No 10 in the brigade – an auxiliary – in the days when the Local Government Board operated. The other two men still alive are Mr J. W. Curtis (the Cromer fruiterer) and Mr James M. Grimble (a retired postman). When Mr Kirby first joined up the only appliance used was a hose reel with no pumping apparatus. The first steamer was introduced in 1886 and the first fire station was in Mount Street. Later on that part of the Town Hall now used as dressing-rooms was the fire station and it was subsequently moved to what is now Messrs A. H. Fox's warehouse in Canada Road. Soon after, in 1905, the present Fire Station was brought in use and Mr Kirby has been in charge as Superintendent ever since. He has served under 5 chiefs, i.e. Hon. Capt. B. Bond Cabbell, Superintendent E. R. Priest, Captains W. G. Sandford, J. K. Frost and T. L. Randall (the present Chief Officer). Superintendent Kirby is full of admiration at the improvements made for fire-fighting and the up to date methods of calling the Brigade. Gone are the days when there was a long wait for horses or a lorry. As in other places, as soon as the fire alarm is given at Cromer, a bell rings in the houses of firemen at once.

Mr Kirby has the Fire Brigade Silver Medal for twenty years' service and five five-year service bars.

The Cromer Brigade with its zealous officers and men is very up to date and it is a proud fact that whilst the men are smart and keen the local council have always taken a big interest in the welfare of their firemen by supporting their needs for good appliances.

Superintendent Kirby has many reminiscences to tell, and while he obliged our Cromer correspondent with two or three, he said, with a twinkle in his eye, he preferred to leave the others until that hardy annual, the Cromer firemen's social next month. He remembers going to one fire in his very early days and removing some curios from a store with the owner looking on considerably perturbed. Only when all the curios had been removed from the smoke and flames did the owner divulge that the curios were filled with live cartridges!

At a more recent fire, with Captain T. Randall in charge, the lorry was pulling home the steamer from a fire outbreak in Northrepps, when two of the firemen complained of hot feet. Until two women called attention to it, the firemen did not know that the lorry was on fire. The outbreak was extinguished, but Mr Kirby did not say if it counted as a service.

Mr Kirby's name is not only associated with a large and respected family of Kirbys in North Norfolk, but the older generation of ambulance men and Life Saving Apparatus Company will recall his association with those voluntary institutions. For twenty-eight years he was with the latter and he is the proud possessor of a medal for twenty years' service, together with a letter of appreciation from the Board of Trade.

Mr & Mrs J. Kirby stand by the rear door of the Fire Station.

Minute books 1918–39

Study of the minute books has revealed certain key themes, which we have isolated and shown in sequence above. There were, however, many other interesting issues which developed slowly over a considerable period of time, or were of short-lived relevance only. These follow now in a presentation through summary.

December 1918
Southrepps opts out of arrangements for using services of brigade. Mr Curtis re-elected as Chairman. Arrangement of 2 youths sleeping at station to continue.

December 1919
Chief Officer warns that, after 33 years' service, the condition of the steamer is poor. Replacement needed in near future. Tenders for 400 ft of hose.

December 1920
Estimates sought for repair or replacement of steamer. Aldborough want to join parish arrangements. Mr Palmer wants to give up horsing of fire engine. Mr Wilkin willing to take over.

The steamer in action in West Street. The fireman on the right is Billy Lusher and nearer the engine is Bob Balls.

January 1921
Examination of steamer reveals it is impractical to repair. Quotations sought for new or second-hand steamer. Mr A. H. Fox resigns as 1st Lieutenant.

December 1921
Second-hand steamer bought from Merryweather. Old one offered for sale. First mention of motor transport for steamer. Death of John Hayward Kettle, fireman since beginning of Brigade.

May 1923
Old steamer sold for £15 10s.

January 1924
Additional storage arranged with football club. Telephone link between Suffield Park and Fire Station sought.

November 1924
Ambulance lectures arranged for members of the brigade.

May 1925
First mention of chemical extinguishers.

September 1925
Emergency station at Suffield Park – sufficient water pressure?

November 1926
Sale of 100 ft rubber-lined hose to Runton Parish Council for £10. Roll map of hydrant positions purchased. Purchase of 2 x 50ft lengths of Thistle brand canvas hose with instantaneous couplings.

February 1927
Permission granted for annual brigade Smoking Concert to be held at Fire Station. Proposal for Runton's fee to be increased from £7 10s to £10 10s.

May 1927
Proposal to increase number of firemen from eight to ten. Runton agree increase of fee.

December 1929
Great satisfaction expressed at new fire engine. Fires attended at Wolterton Hall, Royal Links Hotel and the Buildings, Trimingham. Letter from Guy Davey, agent to the Wolterton estate, expressing congratulations and thanks.

Captain Frost

Captain J. K. Frost was involved from the very beginning of the brigade, eventually resigning on 11th January 1924 after a period of non-stop service approaching 40 years. The minute-book, under 11th January 1924, reads:

The Clerk as Hon. Captain stated that he desired to resign his appointment but would be willing to continue his services provided he was released from direct responsibility for the working of the Brigade by the appointment of a Brigade Captain. It was resolved to recommend:

1 that the resignation of Mr Frost be accepted with great regret and appreciation of his long services.

2 that Mr Frost be re-appointed Hon. Captain of the brigade on the condition quoted above.

3 that Mr T. L. Randall the present 1st Lieutenant be appointed Brigade Captain

4 that Mr Victor Bullen, the present 2nd Lieutenant, be appointed 1st Lieutenant

5 that Mr E. W. Harrison of 33 Bernard Road be appointed 2nd Lieutenant.

6 that the Committee be empowered to make such alterations in the Rules and Regulations of the Brigade as may be necessary to meet the altered conditions.

This last clause leads us to think that the Committee were perhaps taken aback by Mr Frost's announcement, and rushed into making new provisions. If in so doing they broke some of the existing rules, then they would change those at the same time!

We may imagine that for the final months of his service Mr Frost was a sick man, unwilling to renounce his active role in the brigade. The sad fact is he was to die less than ten weeks later.

James King Frost was a much respected member of the Cromer Community, to the extent that most business premises closed on the occasion of his funeral on 17th March 1924. He was 64 when he died. The then Captain Theo Randall and Lieutenants Bullen and Harrison led the cortège and brigade members acted as bearers.

A resident of Cliff Avenue, Mr Frost was Clerk to Cromer Urban District Council at the time of his death, a post he had held since 1901. A very capable man when it came to paperwork, he was also active with the Cromer Protection Commissioners (the body responsible for looking after the sea defences) and was a Deputy Clerk at the Magistrates court. He also used his talents in support of the golf club and the football club.

He was on the staff of Messrs Hansells and Hales, solicitors at Cromer; the Council's tribute in his funeral report describes him as 'an honourable and worthy servant of the public and a true personal friend'.

June 1929
First mention of smoke helmets. Brigade currently has one; request for a second granted. Fireman C. Swann had committed breaches of discipline; ordered to leave service.

December 1929
Discussion of raising fee charged to neighbouring parishes for services of brigade. Decision deferred. Letter from railway company refusing to accept liability for brigade's expenses after their attendance at Roughton Road signal box.

March 1930
Superintendent lodges complaint that there remains a damp problem with the outside walls to his accommodation; help, although promised, has not been forthcoming. Appointment of Dr Colvin-Smith as Honorary Surgeon to the brigade confirmed.

December 1930
Purchase of new equipment: two new electric torches, one pair of gauntlets and goggles for motor engine driver, one megaphone.

September 1931
First mention of possible nationalisation of Fire Service. Engineer Swann refuses permission to Post Office engineer to enter his home in connection with defunct fire bell. Chief Officer Randall sorts out problem.

BACK ROW: *P. Hastings, C. J. Swann, T. Allen, J. Everitt, W. J. Lusher, S. Nockels, J. Danaher.*

MIDDLE ROW: *C. Randell, G. Crane, W. Hawes, E. Kirby, R. Balls, C. Durrant.*

SEATED: *A. Royall, J. Kirby, T. Randall, S. Bastow, A. E. Willins, H. Bacon, E. W. Harrison, V. Bullen, F. Smith.*

Inter-Brigade Competitions

RIGHT: *Theo Randall taking part in one of the Inter-Brigade Competitions. The object of the exercise was to hit the bucket of water suspended from the framework and get away before getting wet!*

We have already seen that relations between fire brigades of neighbouring towns were on an informal basis, at best. There was some willingness to co-operate, but issues of compatibility of equipment and rules of engagement had not been addressed at an official level. Two initiatives got under way in the mid 1930s in a bid to bring more effective arrangements into being. The first was through a Home Office bid to secure the 'co-operation of Fire Brigades of North Norfolk', and the second was a plan to set up Inter-Brigade Competitions. It has to be said that progress on the first was slow; a 'draft scheme' gave way to a 'revised scheme'; decisions were deferred; further clarifications were sought. The language used in the minute-book shows a degree of trepidation:

whilst in sympathy with the scheme the Home Office be asked to supply further particulars of its financial provisions before coming to a decision on the matter.

Long-held autonomy and civic pride were under threat here – the Home Office bid was getting nowhere.

Enter Mr T. A. Cook, MP, a man who enjoyed the high profile that being a Member of Parliament allowed. He was to prove decisive in both areas of development. He wrote a letter to the Committee which gave them the necessary reassurance about entering the scheme, and he encouraged the inauguration of Inter-Brigade Competitions by offering a venue at his home, Sennowe Park.

3rd September 1934 *The Clerk read a letter from Mr T. A. Cook MP, stating that at a meeting of the Norfolk Fire Brigades' (Competition) Association it had been suggested that at Whitsuntide 1935, a fire brigade camp should take place at Sennowe Park from Saturday June 8th until Monday June 10th for the purpose of general instruction, lectures, competitions etc. for the promotion of improved efficiency in firefighting.*

The competitions took place over the next five years, preparing the way for the very necessary co-operation during the Second World War.

LEFT: *The Cromer fire engine in front of Felbrigg Hall.*

December 1931

Mr Ketton-Cremer of Felbrigg Hall told that Fire Service is prepared to hold drill at hall for demonstration purposes. Roughton, Felbrigg and Aylmerton are invited to join arrangements for use of Fire Service if needed.

BELOW: *On parade in Louden Road in the 1930s. They are all wearing their medals, so perhaps this was for a Remembrance Day parade?*

This photograph was taken in front of the Fire Station by Harry Royall on 2nd January 1932.

BACK ROW FROM LEFT: *C.Swann, A. Royall, R. Balls, G. Crane, H. Mitchell.*

MIDDLE ROW: *F. Smith, (large gap), C. Swann Sen., E. Kirby, J. Danaher, P. Hastings.*

FRONT ROW: *T. Allen, J. Everitt, C. Durrant, T. Randall, J. Rounce, J. Kirby, E. Harrison, H. Cooper, S. Nockels, W. Hawes, C. Randell, W. Lusher.*

FRONT ON LEFT: *(standing) T. Stonell, (gap), (seated) Mr Crome, Mr Gowing.*

March 1932
Captain and First Lieutenant complain that the Fire Alarm bells installed in their homes do not ring 'loudly or distinctly enough'. No answer as yet from above parishes.

June 1932
Chains for rear wheels of motor engine are recommended. Above parishes join agreement.

June 1933
First mention of inter-brigade competitions (suggestion by Chief Officer Randall).

September 1933
Brigade had taken part in Hospital Carnival. Five local brigades accept invitations to take part in competitions.

November 1933
Special meeting called to discuss official moves for bringing the fire brigades of local towns together.

December 1933
Cromer appoints Mr J. J. Kemp an 'Honorary Instructor' (i.e. coach) to help prepare the brigade for entry into competitions. Report on a preliminary meeting in Holt on the question of 'Co-operation of Brigades'.

January 1934
Second Lieutenant E. W. Harrison appointed 1st Lieutenant. Cromer Hall Estate disclaims liability for services of Fire Brigade in attending a fire near its woods. Chief Officer's comment: 'but for the services of the brigade the East Wood would in all probability have been destroyed.'

March 1934

Home Office suggests a Joint Committee be set up to advance 'Co-operation of Brigades.' Home Office offers various equipment for trial. 'A new type of portable reviving apparatus for use on persons apparently drowned or asphyxiated' is of interest.

March 1934

Norfolk Fire Brigades (Competition) Association formed. Cromer offers to host the first competition on the Meadow, on 1st June.

June 1934

Report on recent competition: no prizes for home team! Some old leather hose still in use; proposal to replace with canvas.

June 1934

Letter from Mr T. A. Cook, MP, in connection with 'Co-operation of Brigades'. Ongoing correspondence with Cromer Hall Estate regarding non-payment of monies.

June 1934

Circular received about King's Police Medal, to be awarded for 'exceptional courage' or 'conspicuous devotion to duty'. Portable reviving equipment now on trial. Police being instructed on its use. Available on beach during summer?

BELOW: *Firemen photographed by H. H. Tansley in 1935. Left to right: George Crane, Bob Balls, J. Everitt, A. Royall, Commander Harrison, C. Swann?, Superintendent Theo Randall, Harry Cooper (from East Runton), Baldro, Cecil 'Ducky' Swann, C. Randell.*

Putting up the bunting (thought to be for the Silver Jubilee).
ABOVE: *in front of Barclays Bank (picture by H. H. Tansley).*
BELOW: *Next to East House, which was destroyed by a wartime bomb (see page 53) (picture by Philip Vicary).*

September 1934
Conference of National Fire Brigades Association to be held in Cromer.

December, 1934
Suggestion received that the motor fire engine should be fitted with a windscreen. Suggestion that the fire escape should be able to reach the 'topmost storey of any building in Cromer'. Present escape inadequate. Inspection of hotel premises ordered.

4th March 1935
Committee receives offer from a Mr Reynolds to purchase his fire engine for £50. Fire escape could be mounted on this machine, for a further £25.

5th March 1935
Special meeting convened to discuss Mr Reynolds's offer. Agreement reached to purchase his machine for £35.

December 1935
Inspection of hotel premises reports that new fire escape is able to reach the height required. Challenge Cup initiated in Norfolk Fire Brigades (Competition) Association.

March 1936
Decision taken to sell old steam fire engine.

June 1936
Winners! The Cromer Brigade won the Norfolk Fire Brigades' Association Championship Cup. Harry Mitchell appointed as Auxiliary Fireman.

June 1936
Demonstration of old steam fire engine for possible purchase by Mundesley Parish Council. Invitation to the Norfolk Fire Brigades' Association to hold the 1937 competition at Cromer.

September 1936
Proposed increase of fees to outside parishes. Sale of old harness, formerly used in connection with hauling old steam engine, for £1 5s.

September 1936
Proposal to replace six canvas buckets.

December 1936
Laying of sewer in Canada Road. Motor fire engines to be housed at premises of East Coast Motor Company in the interim. Invitation from Mr T. A. Cook for Cromer to attend camp at Sennowe Park in May 1937.

March 1937
Home Office circular about measures to adopt in event of enemy incendiary attack.

June 1937
Second place to Cromer in annual Norfolk Fire Brigades' Association Championship Cup. Decision deferred on above measures until clarification of financial issues is received.

June 1937
Mr J. J. Kemp, Honorary Instructor, resigns his post. Use of station authorised for lectures in connection with air raid precautions.

September 1937
Telephone extensions to homes of Chief Officer and Superintendent approved. Agreement in principle to double strength of brigade, with aid of Government finance.

December 1937
New regulations on vehicle lighting prompts update for motor fire engine.

March 1938
Continuing dispute with Mr Denham Spurrell of Bessingham over payment for attendance of Cromer Brigade at stack fire. Legal proceedings instituted.

May 1938
Cromer win Bevir Cup competition for surprise turn-outs. Steam whistle of Cromer Laundry to be used for daytime fire calls.

December 1938
Informal arrangements of mutual co-operation between Cromer, Holt and North Walsham to continue.

December 1938
ARP auxiliaries now receiving training. Clerk reports on the main provisions of the Fire Brigades Act 1938.

Events and incidents

Roughton Heath 20TH OCTOBER 1927
Gorse fire near Isolation Hospital. Call by messenger. Six men taken by cab from East Coast Garage and fire beaten out ten yards from hospital fence.

Cromer Church 10TH NOVEMBER 1927
Call to fire caused by lightning bursting gas main. Engine not required. Emergency hose cart taken. Fire subdued with wet earth and two men left on duty all night.

Roughton Road signal box 14TH JULY 1928
Call by phone from M. G. & N. station: signal box in danger. Fireman Swann sent on motorbike from station to Overstrand and message received at 4.15 p.m. Captain and Lt. Harrison proceeded by car and found fire on railway banks and in woods but not sufficiently near for signal box to be in danger.

Royal Links Hotel garage 24TH NOVEMBER 1928
Call received by phone 3.45 a.m. Motor engine at scene of fire 12 minutes later. Garage blazing and flames fanned by strong westerly breeze. Started pumping at once with two lines of hose, 500 ft each. Pumped for three hours straight away and afterwards intermittently. Fire extinguished by 8 a.m., but relays of four men left on duty with hose direct to hydrant until 2.45 p.m.

East Coast Garage 12TH JANUARY 1929
Call by phone at 6.30 p.m. Engine manned and at scene of fire at 6.35 p.m. Started pumping at 6.39 p.m.

Hospital ceremony 16TH OCTOBER 1930
Officers and men as per presence book attended at the Foundation Stone Ceremony at the new Cromer and District Hospital.

Carnival display 26TH AUGUST 1931
Fire Display at Cromer Hall with motor engine and escape.

Mill Farm, Thurgarton 22ND AUGUST 1934
Call at 2.55 by phone from Mr H. J. Wright, tenant of farm. Arrived at fire at 3.20. Farm buildings and haystack on fire but no further danger to adjacent house. Found water half a mile away and sent back for more hose, ran out 1850 ft hose, but still short. Sent Lt. Rounce to Sheringham to borrow 750 ft hose, connected up and pumped for one and a quarter hours on stack and buildings. Lt Rounce 2 cars, refreshments 15/-.

Fighting the Weybourne plantation fire.

Weybourne 10TH AUGUST 1935
Call by Mr Howes (agent to Lord Walpole) to woods above the Weybourne Springs Hotel. Very extensive plantation fire. Assisted by Holt Brigade. Pumped from springs to top of hill by relay and checked fire on crest. Moved to woods near hotel and pumped on margin and undergrowth near wood huts; stood by all night and left all clear. Refreshments Captain 10/6d Lt Rounce 16/- Superintendent Kirby 3/-. Hire of car 10/-. 2 messengers.

Manor Farm, Bessingham 23RD AUGUST 1937
Another dispute over payment.

S. F. Gee's cycle shop, Bond Street 28TH DECEMBER 1938
Called by phone. Shop and stores well alight. Pumped from hydrant top of Bond Street. Extinguished fire and left all safe.

Quenching the flames at Gee's cycle shop in Bond Street.

Captain Randall reminded his audience that in 1940 vehicles with solid tyres would no longer be allowed on public roads. The brigade's escape vehicle had such tyres and a replacement unit might be necessary.

Preparations for war

Nearly two and a half years before the event, the nation was getting prepared.

1st March 1937 *The Clerk reported the receipt of a circular and memorandum from the Home Office with reference to Emergency Fire Brigade Organisation and the strengthening of brigades by the provision of additional appliances and the recruitment of additional personnel as a safeguard against the risk of incendiary attack from the air.*

Six months later, the Council approved in principle 'the doubling of the strength of the brigade by the training and enrolment of Reserve and Auxiliary Officers and Firemen and the provision of certain additional equipment with the aid of Government financial assistance'. In March 1938, it was reported that the brigade had been attending weekly lectures on air raid precautions, and in December of that year they received a Light Trailer Pump courtesy of the Home Office, one of whose representatives also made the suggestion that a suction pump be installed at the end of the pier as a supplementary water supply in an emergency.

Other moves in preparation for war included the provision of a shed for the purpose of physical training and an attempt to obtain grant aid in order to extend the fire station.

Another smart line-up of the 1930s Brigade.

World War 2

As we have seen, there had been considerable preparation for the war which was declared on 3rd September 1939. At government level, moves had been afoot since 1935, but the local response had been rather patchy. For example, Cromer Water Committee felt it too unsightly, war or no war, for the water hydrants to be made more conspicuous with a coat of paint, and therefore declined to grant permission. In similar mood, the Council refused to provide a second siren at Suffield Park. Air-raid shelters had been acquired and some trenches had been dug in North Lodge Park and Cadogan Road car park, but sufficient for only 10% of the population. A first-aid post was established in the basement of the Grand Hotel. As regards evacuees, the town was told to brace itself for an influx of four thousand. The beaches were barb-wired, the town was overrun by troops and most of the hotels commandeered. And in July 1940, at the height of anxiety over a possible German invasion, the unthinkable was ordered – to blow out a gap in the pier with explosives to deter any possible landing force.

As regards the Fire Service, we have seen that moves had been made at local level throughout the thirties to improve levels of efficiency in individual brigades; further, first steps had been taken to develop a co-ordinated response to an incident. The Home Office had issued circulars, to which Cromer Fire Brigade had initially made a somewhat guarded response. After all, it had been operating as an individual unit for over 50 years, and the move to greater effectiveness in operation entailed some loss of autonomy. However, one of the major problems which had plagued the service ever since its establishment – the issue of payment for attendance at fires – had been solved by the passing of the Fire Brigades Act in 1938. Section 5 of the Act ended the close association between fire insurance companies and firefighting by abolishing the right to charge for services. Overall, the Act made it a legal requirement for borough and district councils to make adequate arrangements for an efficient fire service. The Home Secretary was also given various powers to provide some Central Government control.

Cromer's full-time firemen during the war years
BACK ROW, LEFT TO RIGHT: *George Harrison, Alf Martin, Jimmy Dennis and Aubrey Thompson.*
MIDDLE ROW: *Jimmy Craske, A. Kirby, Eric Kirby, Edward Girling Smith, Harry Mitchell and Arthur White.*
FRONT ROW: *Sammy Nockels, Phyllis Frost, Column Officer Trollope, Theo Randall, Stella Kemp and Catherine Bird.*

With that background, Cromer Fire Brigade prepared to defend the civilian population from the ravages of fire brought by acts of war. Further on in these pages we identify many of the individual steps which all contributed to this underlying aim; now we may chart the main areas of development. The main station in Canada Road was supplemented by a sub-station in Suffield Park. The Home Office supplied new equipment, including light trailer pumps. There was a steady increase in personnel, with the appointment of full-time firemen, the stepping up of the Auxiliary Fire Service, the engagement of telephonists. In a very informative newspaper article, Tony Rogers writes:

The sub-station at Suffield Park Hotel was manned by Aubrey Thompson and Eric Kirby. The full-time men were on call twenty-four hours a day and would stay at the station. During the night watches, the firemen slept in a house in West Street behind the station. Telephonists and catering staff had quarters above the station and all the meals were cooked by Ida Amis. Others who did telephonist duties were schoolmaster Bill Askew, Sidney Cook, journalist, and housewives Ellen Balls, Agnes Ford, and Ivy Alexander.

They mobilised the best vantage point in town.

One of the duties of the part-timers, trained by the full-time men, was fire-watching from Cromer's highest point, the church tower. They had a hut up there for comfort, but on one occasion the hut caught fire and the full-time men had to be called out. They arrived to find Rev. Gilbert Barclay climbing the stairs with a fire extinguisher to put out the flames. The firemen had to use a hydrant near Barclays bank and linked lengths of hosepipe together to douse the fire. The remains of the hut were thrown into the churchyard.

Ruby Fisher was working as a telephonist in the fire station the day war was declared. In a recent conversation, she gave us intriguing insights into life in Cromer in these early months.

Ruby Fisher *I was on duty the Sunday morning, September 3rd, when war was declared. I took the Red Alert. The Kirbys were living upstairs; my work station was downstairs, next to the fire engine. All that afternoon people were filling sandbags and putting sticky strips down their windows, hanging blackout curtains. And when I went into work at the fire station the next day, we were well sandbagged up and there was not a speck of light to be seen at all.*

That was a really swift response.

Ruby Fisher *It was – people had been attending classes for months and months because of the threat of war, so they knew what to do.*

Had you been working there as a telephonist long?

Ruby Fisher *I volunteered in 1938, one of three telephonists doing different shifts. At that stage I was working in millinery and window-dressing at Balcony House, where I'd done a two-year apprenticeship after leaving school in 1928. My employers were very good; if the siren went, I had to down tools, grab my gas mask and my tin hat and run up to the fire station. I always had my AFS badge on, and my identity disc.*

Ruby married in 1940, and joined her new husband on his farm in Gunton Park. Back in Cromer one evening . . .

Ruby Fisher *I remember there was a sentry post at the top of Chapel Street, and my husband and I had been to visit some friends on the Runton Road. If you were out and about later than ten o'clock, you were stopped; this voice said, 'Halt! Are you friend or foe?' We said 'Friend' and he said 'Advance friend and be recognised.' Well, we showed our identity cards and we were allowed through – my parents lived in Mount Street.*

Ruby Fisher

Rapid developments

The first set of minutes to which we have access for the war period shows a marked increase in pace in effecting change. For one, the sea was now available in the fight against fire.

4th March 1940 *The Chief Officer reported that the suction pipe at the end of the pier had recently been completed and found satisfactory after a test pumping to the top of the cliff.*

The drive to improve coordination between individual brigades moved from the local to the regional plane.

4th March 1940 *It was reported that in conformity with Home Office requirements, a Regional Exercise to try out the communications between the brigades in the district had been held on February 22nd.*

The welfare consequences of having more personnel available for longer periods of time were being considered.

4th March 1940 *A circular was received from the Home Office authorising the supply of food to personnel when engaged in Fire Service for long periods and the stocking of non-perishable food suitable for use in emergency rations.*

The brigade had to keep pace with the development of new technology.

4th March 1940 *Consideration was given to the question of changing the system of operating the Fire Alarm Bells to the battery system operated by press buttons thereby saving time in calling the brigade.*

Small-scale transport was an issue. The old Chrysler needed replacing, but the funds were sufficient to cover the cost of only a seven-year-old vehicle.

4th March 1940 *An offer was received from the East Coast Motor Company Limited to supply a second-hand 12 cwt van of 1933 make in good mechanical condition at the rate of £17 10s 0d.*

Terms were concluded for the use of premises for the sub-station in Suffield Park.

4th March 1940 *A letter was received from Mr Carpenter accepting the offer of £30 per annum subject to three months' notice for hire of accommodation in the yard of the Suffield Park Hotel for the Suffield Park Sub-Fire Station.*

The higher number of staff involved with fire fighting created the need for further accommodation close by.

4th March 1940 *A letter was received from Mr Tovell with reference to the use of a room at the White Horse Inn as a Rest Room for the Auxiliary Fire*

Service and of a shed in the Inn yard, suggesting a rent of £26 per annum including lighting.

Vehicular access to the Canada Road station was proving difficult.

4th March 1940 *It was reported that the crossing to the Fire Station in Canada Road was very difficult to negotiate with a trailer pump towed behind a car by reason of the angle to the approach to the road and the camber on the surface. The Surveyor submitted an estimate of £5 for altering the crossing to give easier access.*

Working together: the first big test

Nearly a year into the war, a major incident provided the test for the effectiveness of local brigades combining into one coordinated force. The minute-book reports:

2nd September 1940 *It was reported that on the night of 25th–26th August a fire had been caused apparently by enemy action at the Maltings, Great Ryburgh. As the Fakenham Brigade had been unable to cope with the outbreak, other brigades from Cromer, Reepham, Wells and Sheringham had been called and the Chief Officer as District Officer under the Regional Scheme had been put in charge. The fire had been difficult to extinguish and it was not until Saturday 31st August that the Cromer Brigade had been dispensed with leaving the Fakenham Brigade to watch the fire and deal with small outbreaks. It had been necessary to pump continuously for 165 hours.*

It should be noted that a coordinated approach involved considerable adjustment for brigades in terms of chain of command. Under the new arrangements, officers had to learn to receive orders from the senior man on the spot, where previously they were in sole charge of an operation. Another issue of this first test case seems to have been the financial arrangements; the Home Office was reluctant to pay Union rates for the services provided!

Co-operation with the military

An entry in the minutes late in 1940 reminds us of the overall situation in North Norfolk in the early days of the war. North Norfolk became a militarised area, with forces encamped throughout the region. The reason for this was fear of invasion, and the fact that one of the possible sites on the German list was Weybourne, where there is deep water close to the shore. But relationships between the various forces charged with protection of the public were not always straightforward. Here is a sample of how the Cromer Fire Brigade insisted on its rights, keeping the mere army in its place.

2nd September 1940 *The Chief Officer reported that he had been in touch with the Military Authorities with reference to co-operating with them in dealing with fires. The military had an adequate supply of hose but had been unable to obtain standpipes to fit the Cromer hydrants and he had therefore lent them 3 standpipes at the Marlborough. Military fire parties were being organised and trained but they had been informed that they would not be able to use the Council's hydrants without permission and then under supervision.*

A further problem is highlighted here.

4th November 1940 *The Chief Officer reported that he had experienced difficulty in co-operating with the Military Authorities for the provision of fire parties owing to the frequent changes in units stationed at Cromer. He was in communication with the officers commanding the units now in Cromer.*

However, no solution is in sight.

2nd December 1940 *The Chief Officer reported that owing to changes in the units stationed at Cromer he had been unable to progress much further in co-operating with the Military Authorities to secure protection from fires.*

Further developments

Since its inception, the Fire Brigade Committee had met two or three times a year. Now it was felt to be urgent that it should meet more frequently.

3rd June 1940 *It was resolved that for the present the Committee meet every month on the same evening as the Sanitary Committee with the exception of the evening when the Burial Board Committee met.*

The next item (transcribed word for word) caused so much worry to the writer of the minutes that his/her grammar and syntax came under severe stress!

3rd June 1940 *A confidential circular was received from the Home Office Fire Brigades division pointing out that desirability that an auxiliary fireman should be allotted for sentry at the station entrance in order to prevent the entry of unauthorised persons as far as practicable should be armed at the more important stations.*

However, there was no great rush to fill this particular role.

2nd September 1940 *The Chief Officer reported that he had been unable to obtain the services of the Home Guard for the guarding of the Fire Sta-*

tion, but the stations were never left unattended and he had lent an automatic pistol for use if necessary.

There was, nevertheless, a 24-four hour presence in Canada Road.

2nd September 1940 *Arrangements had been made for 5 to 7 of the AFS personnel always to sleep in the station, the remainder turning out only when specially called or if bombs dropped in the district; the necessary beds and mattresses had been purchased. A gift of £50 had been made for the benefit of the Cromer Civil Defence Services, and the Cromer Fire Brigade had been allocated £2 10s which had been used for the purchase of pillows.*

Provision was made for staff in the event of attack from the air.

2nd September 1940 *The Chief Officer reported that air raid shelters for personnel had been provided in the yard of the White Horse and in the basement of the Suffield Park Hotel.*

However, quality of materials could not be assured.

2nd December 1940 *Authority was sought for the purchase locally of green wood to strut up the sides of the air raid shelter for personnel in the White Horse yard which had collapsed at one point. The price of such timber was very reasonable.*

Issues of training were not neglected.

2nd December 1940 *Attention was drawn to the arrangement which had been made by the Ministry of Home Security whereby ARP training films could be hired at 5/- per day and the Chief Officer asked for authority to hire films dealing with incendiary bombs and with rescue and first aid work. Mr H. H. Tansley had been approached and was prepared to show the films through his projector at a reasonable fee.*

All the time, a steady mobilisation of personnel was continuing; those directly involved with the Cromer Fire Brigade now numbered 17 in the regular brigade, 37 in the AFS, with 11 telephonists and four messengers. There was also scope for other civilians to play their part.

3rd June 1940 *Circulars were received from the Home Office with reference to the formation of supplementary fire parties composed of selected members of the public who could be trained in the use of stirrup hand pumps. The number of pumps originally allocated to Cromer if the Council co-operated in the scheme was two and this had now been increased to five. The Home Office had been informed that five supplementary fire parties could be formed in Cromer and the issue of the necessary stirrup hand pumps was awaited.*

Fire party training was not neglected.

3rd March 1941 *A specimen handbook for the use of Fire Parties was submitted and the Chief Officer stated that in his opinion the book would prove to be very useful.*

But only the very fit could join the war effort in the following key location.

3rd March 1941 *It was reported that the scheme for spotting fires from the Church tower was now on a satisfactory financial basis; 'fire spotters' were now employed and had been enrolled as Air Raid Wardens.*

Cromer church tower made an excellent vantage point for watching the whole town for fires.

Incidents in the early part of the war

We have here been seeing the various measures undertaken to face the onslaught when it came. Fortunately, direct attacks were few, although a bomb landing in Central Road had a devastating effect, with several civilian deaths. In his excellent book *Coastal Towns at War*, Peter Brooks writes:

A much more serious raid took place at 6.35 a.m. on Sunday, November 17th when five high explosive bombs, plus several incendiaries and one unexploded bomb were dropped on Central Road and Suffield Park. Despite touring the town, the Fire Brigade found no fires; any that might have been started, especially in the Grove Road, Hospital area having already been extinguished by the ARP services. Central Road presented a more horrific scene. Number 18 had received a direct hit, a bomb passing directly through the bed in which 12 year old Doris Emma King was sleeping before passing through to the ground floor and exploding. Other occupants of this small terrace house were Mr and Mrs King and Mrs King's mother, Mrs Hilda Allen. Mrs King recalls that all three of them knew nothing of the attack until they woke up to find their faces covered with bits of plaster and the sky visible above them.

Elsewhere in the country, however, there were sustained bombing attacks, and Cromer Fire Brigade found itself involved in the national scene. As part of a further coordination in firefighting work, brigades from the local towns were likely to receive a 'Regional Call' summoning them to be on stand-by in case of need. So it was they went to Cambridge, Yarmouth, Norwich, Chesterfield, Nottingham and Romford.

In hindsight, it is possible to see the above as part of much wider scheme, to create a fire service which could properly serve the whole nation. We have seen from the very beginning the moves at local level to bring individual brigades into joint operations. The logic to this development led to the formulation of plans for a National Fire Service, and the first formal hint of this came in the early summer of 1941.

3rd June 1941 *The clerk reported the receipt of a circular dated May 13th from the Home Office with reference to the re-organisation of the Fire Services during the present emergency, stating that the whole of the Regular and Auxiliary Fire Brigade resources of the country would be placed under the general control of the Secretary of State.*

The Fire Services (Emergency Provisions) Act came into force on 1st July 1941, with responsibility for the finance and direction of the Fire Service passing out of local hands. The net effect on Cromer was that the brigade lost its autonomy and became one element in a network of provision, while its chief, Theo Randall, became responsible for a large area of North Norfolk, stretching from Wells to Stalham.

SS Teddington

One of the major incidents to involve the Cromer Fire Service in the first part of the war was offshore; it tested to the very limit the coordinated efforts of firemen from Cromer and Sheringham, the crew of the lifeboat and salvage teams sent from Yarmouth. A convoy steaming up and down the coast came under U-boat attack, and two of the German submarines closed in on one of the merchant vessels, the SS *Teddington*. They both fired a torpedo; the ship was struck, set on fire and immobilised.

Nearly 60 years later, the full story came to light in a remarkable way, involving great diligence, determined research efforts on both sides of the North Sea and a degree of luck. In the year 2000, the daughter of the first mate of the *Teddington*, Mrs Susan Ashenden, called at the Lifeboat Museum with an enquiry about the sinking of her father's ship. Frank Muirhead undertook to help her, knowing that the museum had good working relationships with research teams from Germany. Each party made approaches to the relevant naval archives, and found the necessary documentary evidence. So far, so good. The element of luck came with the contact made with one of the men who had worked on the salvage team, Mr A. I. Knights, who now was a volunteer attendant at Lowestoft Maritime Museum. From memory he was able to provide some key detail which answered some of the questions remaining from research into the documents.

So, what exactly happened? In *Coastal Towns at War*, Peter Brooks gives the following account.

Cromer lifeboat was launched to go to the aid of SS Teddington, *a 10,000 ton cargo ship which had been torpedoed some five miles out to sea. The ship was carrying a valuable cargo of aeroplane accessories and was on fire. The lifeboat took out two light pumps and accompanying firemen on the first day and then two more pumps from Sheringham the following day. Despite their efforts during daylight the fire regained control during the night and it took three days of dedicated firefighting to get the fire under control. The vessel was eventually moved off Overstrand where efforts were made on salvaging the cargo and in April 1942, Cromer NFS received congratulations from Admiral A. R. Dewar, Director of Salvage at the Admiralty, for their work in fighting the fire, which at times, as Swiggy Kemp recalls, was so fierce the sea round the ship was actually boiling.*

Let us now turn to the account Mr Knights wrote down from memory over five decades later.

The Teddington *was hit by a torpedo in convoy somewhere off Smith's Knoll. It caught fire mainly in No. 2 and 3 hatches. A salvage tug, called the* Nesses, *of which I was a crew member, was sent out from Yarmouth to fight that fire. All day we fought with the help of the fire Service personnel; at one time we had it under control, but as it got dusk the Fire Service personnel were taken off and put ashore, as they were not allowed to be on the high seas after dark. Being a salvage ship, we had to remain with the casualty, with a destroyer to cover us from further attacks; obviously, a ship ablaze lit up the whole sky; we did get attacked, but luckily there were no further casualties. We fought that fire all night, but it got out of hand again and it was decided to let the destroyer tow her until she beached off Cromer; then the destroyer put shots into her below the water line, thereby filling up with water and putting the fire out. If I remember rightly, she was about a 10,000 tonner, outward bound to India with general cargo; but amongst her cargo was a consignment of raw nickel consigned to the Bank of India, stowed right at the bottom of No. 4 hold, worth a million and a half pounds. That's what we had to get out.*

That night we never had a glimmer of light anywhere, but it was bright moonlight, the Germans knew she was there and they had a go at machine-gunning us; fortunately there were no casualties.

The Admiralty decided not to risk their salvage vessel the *Nesses* overnight, so accommodation was found in the town for Mr Knights and two colleagues. The role of these three mechanics was to prepare each morning for the arrival on board of a gang of 29 stevedores, who had been brought down from Immingham to work on retrieving the valuable elements in the cargo. They, too, were lodged in the town. However, there was potential danger from the English side, too.

Each day one of your cobles was chartered to go out to the ship at about two or three in the morning, to put the three of us on board. We had to work carefully – and just by the light of torches. Our job was to get the pumps started and the water pumped out of No. 4 hatch by the time the Cromer lifeboat brought out the stevedores. By the way, her hatches were completely covered at high water. Soldiers guarding the shore were ordered, if they saw a coble, not to fire on it, as their target would be we three mechanics and the coble's skipper; thank God they never fired a shot! Sometimes it was a bit rough out there and a bit scary; on one occasion, Henry Blogg came out with the stevedores, and instead of putting them on board he had to rescue us! . . . anyhow, all the nickel was recovered and we went back to our respective jobs. I personally stayed in that salvage company for a further twenty-six years working all over the world.

The other two mechanics were John Atkinson from Carnforth and Arthur Barber from Birmingham.

The minute-books

The minutes chart the gradual evolution of the brigade in terms of equipment, organisation and personnel. We have here made a selection of some of the key matters.

September 1940
Second foam-making branch pipe supplied by Home Office for use in petrol fires or similar.

March 1941
Steel helmets for fire personnel.

November 1941
Recreation room at West Street.

November 1941
Column Officer Trollope appointed. Cromer to be centre of sub-division. Additional accommodation required – modifications to station. Mess-room for about 30 men created out of Kirby's flat.

March 1942
Retirement of over-age full-time firemen, including J. B. Kirby. Static water supplies – tank in the Meadow etc.

June 1942
Purchase of land next to Fire Station considered. Senior Company Officer T. L. Randall appointed Liaison Officer for the National Fire Service to work in conjunction with the Invasion Committee in the Cromer area.

Although no fire engine is to be seen, the pictures on this page and the next give an idea of the level of damage to Cromer from World War 2 bombs. Before the war, East House had occupied the site next to the churchyard (see page 38). It was totally destroyed on 22nd July 1942.

Bomb damage in July 1942 at the corner of Church Street and the High Street.

November 1942
Desirability of appointing Fire Guards to watch for incendiary bombs.
Eight men and 16 women directed to part-time work in Cromer. Total strength of local fire service now over 70.

February 1943
Issue of Theo Randall's pay as Senior Company Officer. Start of building of control room next to Fire Station.

February 1943
Standardisation of fire hydrants. Fire Guard Officer appointed.

March 1943
'Surprise exercise' called by the Fire Force Commander at 1 a.m. Increase of women involved to nine full-time and 21 part-time.

February 1943
Report of Fire Guard Officer – lots of recruits! New control room nearing completion. Area extended to include Mundesley. Column Officer Trollope now based in Norwich. Visit of Deputy Regional Controller.

June 1943
Fire Guard Officer – the zeal of the man! Suggests the job should be full-time.

September 1943
Suggestion taken up – and he gets the job (but not at salary requested: £260,

not £300). Ten whole-time, 74 part-time men, five whole-time, 30 part-time women. Total numbers in Fire Guard 372. National Fire Brigades Association – proposals for post-war organisation.

November 1943
Earlier suggestion of fixed pipe connection from end of pier now being implemented. Parish Hall now being used for training Fire Guards.

February 1944
Fire Guard arrangements 'too cumbersome'! Completion of new pipeline from end of pier.

March 1944
National Fire Service Benevolent Fund established. New fire escape – what to do with the old one? Keep it! Volunteers now number 392.

May 1944
Fire Guard Officer reports there are now 402 volunteers. New equipment has arrived.

5th June 1944
D-Day! Fire-watching from church tower, a volunteer effort, is running out of funds. Continued success in competitions.

4th September 1944
Fire Guard Officer Balls has resigned. Mr James Tribe takes over and suggests helmets are painted according to rank. First sign of relaxation. Continue fire-watching from church tower until end of month.

December 1944
Another sign of relaxation – change in shift-times – reduction of hours on duty.

February 1945
No further action with regard to Fire Guard Plan. Query over post-war re-organisation.

June 1945
Many stand-down measures.

September 1945
Various stand-down measures: recreation hut, use of Air Raid siren for fire alarm,

reduction in number of appliances. Premises in West Street vacated. Tenancy in Suffield Park Hotel terminated. Emergency water installations demolished.

December 1945
Two sets of breathing apparatus. Death of J. B. Kirby.

Incidents of note

The last page of the incident books available to us is dated November 1941, so we have no access to detailed information from the archives after that date. Nor, as we have seen, do the minute-books offer accounts of Fire Brigade response to emergencies. We do know that Cromer suffered intermittently from attacks from the air over the first four years of the war, with the worst raids on 22nd July 1942. Sometimes these were the so-called 'tip and run' raids, in which enemy aircraft unloaded bombs on coastal towns before flying home across the North Sea. On other occasions, they made deliberate attacks. Peter Brooks' *Coastal Towns at War* gives full information here, and from his pages we may quote the following two incidents which include mention of the Fire Service.

The Wings for Victory week 1943
The week of Sunday 13th June opened with King Peter of Yugoslavia taking a mile-long march past of military and defence units. The week was marred by two tragic air crashes. On the opening day, during a firefighting demonstration on the Meadow, a Beaufighter crashed about 500 yards away and burst into flames. The firemen raced to the scene but they could do nothing for the two-man crew who were both killed. Fortunately, there were no other casualties, although there was some damage to nearby houses from flying debris.

Later, after the war in Europe had ended, there came a further tragedy involving a British plane.

Reminders of war still abounded, as when on 25th June 1945, a Halifax bomber from 199 Squadron based at North Creake crashed into the cliff below the coastguard station at Cromer. All the crew were killed, and firemen had to be lowered down the cliff-face to fight the resulting fire. Several people on the beach had narrow escapes, and there is no doubt that many deaths and injuries would have resulted if the crew had not struggled gallantly and successfully to keep the plane airborne long enough to clear the crowds.

Zeal
An account of the Fire Service in Cromer during the Second World War would not be complete without brief mention of very elaborate local plans to

defend the population against fire from the air. After experience of the blitz, the government was keen to see the establishment of a civilian defence force, called the Fire Guard, on a street by street basis. In the report of the minute-book dated 1st February 1943, we read of the following appointment. On your toes, Cromer!

1st February 1943 *Mr W. H. Balls accepted the appointment as Fire Guard Staff Officer on the terms decided by the Council, at the salary of £2 per week.*

A month later he issued the following report.

1st March 1943 *The Fire Guard Officer reported that during the month of February he had made 178 calls with a view to organising Fire Guard Parties and in the Cromer ward 26 persons (18 women and 9 men) had volunteered their services. The planning of the Urban District into sectors and street party areas was proceeding.*

Two months later he had further progress to report.

3rd May 1943 *Since the last meeting 108 persons had agreed to undertake Fire Guard duties in the district, the total number enrolled now standing at 134 (79 women and 55 men).*

By now Mr Balls was getting into his stride. He submitted his plans for the town.

7th June 1943 *The scheme provided for a central organisation consisting of a committee comprising three members, the division of the town into 4 areas corresponding to the existing Wardens Areas with an Area Officer in charge . . . for the areas to be sub-divided into sectors, eight in all, having a Sector Captain in charge; and for the sectors to be sub-divided into streets or blocks of 20 to 30 houses each street or block having a Street Captain in charge with street party leaders as required.*

But that was not all for that particular month.

7th June 1943 *The Fire Guard Officer reported that 341 volunteers had been enrolled . . . and were prepared to undergo training . . . and for this purpose a smoke hut would be of immense value . . . and that to carry out the scheme it would seem necessary for a whole-time Fire Guard Officer to be considered.*

Three months passed.

6th September 1943 *The clerk reported . . . it was necessary for the Council to appoint a whole-time Fire Guard Officer. Mr Balls said he would be willing to accept the appointment at the salary of £300 per annum.*

The Committee withdrew to consider the matter.

6th September 1943 *The Committee resolved to recommend that Mr Balls be appointed at an inclusive salary of £260 per annum.*

Mr Balls declined the post at this figure but resolved to continue on a part-time basis. By November, he had a further blow: it was felt a 'modified scheme' might suit Cromer better. Yet more discouraging news came to Mr Balls in February the following year.

7th February 1944 *It was decided to abandon the system of sectors containing a number of street parties as being too cumbersome for a place like Cromer.*

Undaunted, Mr Balls was reporting an increase in numbers the following month to 392, and still hoping for more. By May, he had topped 400. Perhaps that had been his target all along, because by September the Fire Prevention Officer was reporting a vacancy.

4th September 1944 *A letter was received . . . expressing the regret at the resignation of Mr Balls and the hope that the Council would be able to find a successor.*

Mr Balls was replaced by Mr Tribe from Norwich, who undertook his duties on 15th August that year. We do not know his salary, but we can read of a zeal similar to that of his predecessor. He reported an increase in personnel to 416, and submitted a plan to have helmets painted in accordance with rank. But by the following February, conditions were easing, and 'on the relaxation of Fire Guard duties the painting of the helmets had not been preceded [*sic*] with'.

The saga ends with the note that the 416 had been thanked for volunteering for duty. All that effort!

The post-war period

The archives for the years following the end of the war are comparatively bare, but still some key incidents and events stand out. Peter Brooks in *Coastal Towns at War* recounts concerns from a local politician over costs and administration.

At their meeting on December 11th, Councillor Baker enquired what was to be the future of the Fire Service after the war. If they were not careful, he said, there might still be a National Fire Service. He understood that the local station was costing £100 per week to run, and they must see to it that the Fire Service was handed back to local authorities. The firemen themselves received a justifiable mention when, at a cost of barely 10/- for the wood, they made toys, in their own spare time, worth some £80 for distribution at the annual tea and entertainment in the Fire Station. Some 120 youngsters were entertained and before leaving each one was given a 6d Savings Stamp, a 3d piece, and chocolate (for which the men had given up their coupons) and a present from Father Christmas.

In the event, the provisions of the Fire Services Act, 1947 entailed the return of the Fire Service to local authority control; not, however, to town or district councils but to county councils and county boroughs, in all 148 different brigades.

Theo Randall

Theo Randall's name has been mentioned on numerous occasions, and for good reason. His membership of Cromer Fire Brigade dated back to 1911, and he had been Captain (Station Officer) since 1925. He decided to retire in 1954, bringing to an end a period of association with the brigade of more than 40 years. We are very fortunate to have had the opportunity to talk with his

Captain Randall

Theo Randall was one of the five sons and three daughters of Robert Randall, who had moved from Holt to open his clock, watch and electrical business in the town, which continues today as Randalls Electrical. Because of his father's involvement in the foundation of the St John Division in Cromer, he had learned first aid there and joined the Fire Brigade in 1911. He was also known locally for his prowess at football, being a regular member of the Cromer team – after the First World War he would play for Norwich and the Norfolk team, which he captained for a number of years.

Being also a member of the local reserve for the army, he was called up to the Norfolk Regiment on the outbreak of war, rising to the rank of Quartermaster Sergeant. He was involved in a number of hand to hand encounters in that campaign, and skills of leadership and organisation learned there would stand him in good stead when he got back to Cromer. He was very ill for a while as a result of his army service, but resumed his fire brigade work as soon as possible and was eventually promoted to Captain of the Brigade.

His work for the family firm was primarily the selling and repair of watches, and if you asked him to turn back his shirtsleeves, he would unfailingly have several watches on his arms which he would be testing. Perhaps smoke testing at a fire was all part of the process! His workshop was at the back of his wife's sweet-shop at the Crossways in Cromer, and if called from work, he was conveniently close to the fire station.

He served on the Town Council and the Chamber of Trade, and his sporting interests transferred primarily to golf. He led the Brigade through the years of the Second World War, with all the extra responsibilities that it took on then. To his personal duties was added that of supervising the area from Holkham to Stalham. He retired from active fire service in 1954.

daughter, Jane Leeds, born in 1922, and hear about his work at first hand.

What sort of man was he?

Jane Leeds *He was a very laid back man, nothing seemed to phase him at all. We lived behind what is now Lloyds TSB. My mother had a sweet shop there, my father had a watchmakers/jewellers and there was an electrician's on the other side, all of which belonged to us; we lived in a house above and behind the shop.*

Very handy for Canada Road.

Jane Leeds *Yes, when the call-out came he'd make his way up there and sort them all out; one of the highlights of each week was meeting up with them all in the White Horse after drill night. One of his great friends was Commander Harrison – he was also in the Fire Service – who very tragically was killed in*

a road accident in Hartlepool. A great friend and great fun.

We talked of the role of Sir Thomas Cook in the development of the Fire Service in Norfolk. Jane remembered him well.

Jane Leeds *Sennowe Park! I think he got his knighthood for his support of the Fire Service; I believe he came to Cromer at one stage and I presented a bouquet to his very lovely wife, on the Meadow. My father seemed to think 'he rather fancied his weight'. But that apart, there were tremendous displays at Sennowe Park, which were a huge day out for the families as well.*

Any memories of his role in wartime?

Jane Leeds *We used to get advance warning of anything; what's called a Purple Alert came through to our house, via the telephone. I can remember him going to Holkham Hall to advise on storage of art treasures, not a lot else. But what I do remember is a series of arson attacks in the thirties.*

What, here in Cromer?

A Christmas party at the fire station was an annual event, as Jane Leeds mentions in her interview. Her father Theo Randall is seen on the left of this picture, and Tom Barnes, another of our interviewees, is in the back row, holding his daughter. Ian Alexander, who provided this picture, is over to the right and Ray Mitchell, Harry Mitchell's son, is next but one to Theo Randall. Most of the children in this picture will be in their late 50s or 60s at the time this book is published, and the author will be pleased to hear from you if you'd like to identify yourself!

Theo Randall and colleagues at his retirement. The officer shaking hands with him is his successor, Harry Mitchell.

Jane Leeds *No, it was a number of haystack fires round and about. The brigade put the first one out and found a piece of paper saying 'The Lone Wolf has been here.' A week or two later, a similar sort of thing, 'The Lone Wolf has been again.' And then again, 'The Lone wolf strikes again!'*

Perhaps he was talking of his box of matches there!

Jane Leeds *Perhaps he was! We shouldn't laugh about fire, but those incidents do stick in the mind. As does, actually, the smell of the inside of the Fire Station; a very pleasant smell of leather, I suppose – I can remember that from the Christmas parties I had to run after the war – I was a school teacher, so it was reckoned to be my job.*

We also have the press report of his retirement, which adds more details on his career.

After 42 years as an officer and commander of the Cromer Fire Brigade, Station Officer Theo Randall, who was awarded a BEM for his service, retired at midnight on Friday.

He chose as the occasion of his retirement the annual dinner and social of the Cromer Fire Brigade and at this function on Friday evening, glowing tributes were paid to his efficiency and selfless service, and to the efficiency of the brigade he leaves by Mr W. C. Fulcher, chairman of the Cromer UDC and Chief County Fire Officer W. Collow. He also received the gift of an inscribed silver cigar/cigarette box from the brigade members.

Mr Collow announced during the evening that Sub-officer H. Mitchell was succeeding Mr Randall, and that Leading Fireman A. Royall was being promoted to second-in-command.

Mr Randall joined the Cromer Brigade in odd circumstances, being told he had been appointed lieutenant, told he had to report next day to the station to

Harry Mitchell joined the Fire Brigade in 1936. He went into the Forces during the war and became a full-time fireman in Norwich before returning to Cromer, where he became Sub-Officer in 1952.

be sworn in and that he would have the privilege of buying his own uniform. Ever since 1925 until the present time, Mr Randall had carried the responsibilities of the brigade as its chief officer and in addition was at one time a district officer under the Auxiliary Fire Service and later, under nationalisation, was a senior company officer. When the fire service was returned to local control after the war, Mr Randall agreed to continue in charge of the Cromer Brigade . . . Replying to the remarks made about him and in response to his gift, Mr Randall thanked many people who helped the brigade. He mentioned particularly the employers of firemen who were prepared to permit their men to 'down tools' at a moment's notice, and said without their action there could be no fire brigade. Cromer, said Mr Randall, pioneered the St John organisation in Norfolk by organising the first division in the county and also had the distinction of organising the first fire service in the county. Also it was instrumental in forming the Norfolk Fire Brigades Association, aided by Sir Thomas Cook, and the association's annual competitions did much to raise the standard and efficiency of the Fire Service in Norfolk.

Incidents of note

Royal Links Hotel 31ST JANUARY 1949

A major effect to Cromer's skyline came following the destruction by fire of the Royal Links Hotel, built in 1894 close to the Royal Cromer Golf Club.

Firemen attending the Royal Links Hotel fire: T. Randall, R. Durrant, A. Royall, J. Everitt, K. Fairhead, W. E. Storey, H. Cooper, W. D. Mckenzie, C. G. Seago, J. Thetford, A. White, J. Hammond, A. Alexander, H. Mitchell, W. G. Read, C. S. Walker, N. A. Fisher, R. J. Balls, C. Swann.
Picture by Philip Vicary.

ABOVE: *Cromer firemen at Marsham Common.*

BELOW: *Thirty firemen attended the fire at the covered tennis courts, including Station Office H. Mitchell and Divisional Officer G. R. Dix.*

Marsham Heath 30TH SEPTEMBER 1959

Fire units from a wide area were called to heath fires which had been burning sporadically for several days at Marsham Common.

Newhaven Court covered tennis courts JANUARY 1961

These indoor tennis courts gave Cromer a distinguished place on the tennis maps of the 1920s and 1930s. They were, indeed, the premier indoor tennis courts nationwide during that period. Famous players include the legendary French players Suzanne Lenglen and Jean Borotra. The Docherty brothers, who won the Men's Doubles title at Wimbledon eight times, were also visitors here. The courts went into decline in post-war years, but remained an important facility for Norfolk tennis. The building measured 150 ft x 100 ft and had a metal frame, wooden walls with a roof made of glass and asbestos.

The flames were spotted at 3 a.m., leading to a huge blaze which could be seen for miles around. The courts had been in use the previous evening.

Tom Barnes

Tom Barnes was in the Fire Service between 1950 and the 1980s. In a recent conversation, he was asked how he came to join.

Tom Barnes *When I came to Cromer I joined the Water Board. Duckie Swann was our foreman, and he was a fireman – he'd joined when he was a boy; his father had been a fireman, too. So was Dennis Green, who also worked there. So I joined, too. I went to the station. 'Hallo, brother, sign here, you're in!'*

How did you get the call-out?

Tom Barnes *If I was at home, there was a bell in the house – a very large bell! If at work, the siren sounded. Our HQ was North Lodge Park. We always used to down tools, pile into the works van and drive up to the station. There was very little traffic in those days, no one-way system. The only times we didn't go was when the mains water was turned off; we couldn't leave the town without water!*

Was there still a Superintendent at the station, like before the war?

Tom Barnes *Not as such. But there was a Mr Farrows slept in a flat above the station. He was a retired policeman. If he wanted time off to play bowls, one of the firemen had to stand in for him and take the phone message; it wasn't very practical.*

What aspect of firefighting in the early days stays in your mind?

Tom Barnes *There was no relief – you stayed at an incident until the job was done, nobody came to replace you. But one perk was the food provided by the canteen lady who used to come out from Ingworth; she got to all the fires with her caravan. She must have got a call from HQ, she was always there.*

I suppose Cromer was involved with helping in the 1953 floods.

Tom Barnes *Yes, we did a lot of pumping out then supplying water through water tenders. Bullen's built brick piers for emergency tanks to take the water at Salthouse and Cley, for example, and we kept them topped up for weeks. The householders had to collect their water in buckets.*

Does one particular incident stand out in your memory?

Tom Barnes *Probably the fire at Wolterton Hall. There were flames coming out of windows when we got there, but everybody was already getting the paintings out.*

Could you get inside at all?

Tom Barnes *At first we could. Kenny Fairhead and I were hosing down a panelled room when there was a very loud bang and the water went off; we investigated and found a portion of the stone staircase had fallen on to the landing where we were*

Tom Barnes

Ken Fairhead, Station Officer 1968–75. Ken came to Cromer in 1944 and was the town's signwriter, with his workshops in the yard at the back of the White Horse, conveniently close to the fire station. As with a number of officers who served in the Brigade, he also served with the lifeboat as one of the launching and shore crew.

BACK ROW: *Bob Norton, Joe Pope, Eric Kirby, Tom Barnes.*

FRONT ROW: *Harry Pearce, C. Swann, Harry Mitchell, Alfred Royall, Peter Wickers.*

– we had to get out right quick! So then we had to work from the outside.

What about the supply of water?

Tom Barnes *That was the problem – we had to get water from the lake at the back, it had to be relayed from pump to pump. We were called out in the middle of Friday afternoon, and we had to stay there till eleven on Saturday morning. As I say, we were never relieved in those days. But our canteen lady came as well as the bakers from Aylsham.*

This was in the winter?

It certainly was; there was such a sharp frost at night, as soon as we turned the hose off that froze solid at the nozzle. And we couldn't move the ladders up and down – they had to stay where they were, they were all froze up. One of the coldest nights I can remember: even the cows the next morning came to look at the scene – they looked frozen too! About the only good thing about it was the Walpoles had many people helping them – just as well. They had to snatch the oil paintings down, tear them from the wall.

Harry Mitchell was your Station Officer?

Tom Barnes *He was, he was good. He wouldn't put up with anything from HQ! He worked in the electricity showroom nearby – he'd soon write a letter off to HQ if it were necessary and sort it out. Harry's father was a real Labour man, he'd been on the council.*

Cromer Brigade continued to do well during the competitions at this time.

Tom Barnes *We did. I remember one in particular where Bob Norton got that excited when they blew the whistle that he came out without his helmet on! We lost two or three seconds there – but we still won it.*

Notes from a personal archive

The presentation in this chapter will follow a different pattern from the preceding chapters, for two main reasons: first, we have not had access to official records as we had in earlier decades when the Fire Brigade was the property of the town; and second, it is really too soon after the event to gain the necessary historical perspective. What we do have, however, is a very complete personal archive for these years gathered by Jamie Edghill, and from this we are able to gain an insider's view of the events described. In addition, I recorded many conversations with him, in which he gave his personal testimony, as firefighter, of many of the incidents featured. He is also able to give a unique insight into the way in which the Fire Service in Cromer has been very active in developing a community role in the town, with several aims: to raise awareness of the value of the Fire Service to the town, to alert people to fire prevention measures required in daily life and to raise money for the Firemen's Benevolent Fund.

Any modern Fire Brigade has responsibilities far beyond those in its name. It has to respond to many road traffic accidents, it is summoned when flooding has occurred, it performs rescues off cliffs and up trees. But our primary concern in these pages is with its traditional function in its response to fire, as it is here that its impact is most felt in terms of our environment.

In making our selection from the many incidents between 1960 and the present day, we have been guided by the fact that the buildings of Cromer have sometimes changed through calamity, not choice.

ABOVE AND BELOW: *Newhaven Court after the fire.*

Incidents of the 1960s

Newhaven Court 23RD JANUARY 1963

Jamie joined the service in 1962. It was not long before the brigade was involved in fighting a fire which brought the end to a building off the Norwich Road which had played a prominent role in the town since the early days of the twentieth century. Moreover it happened during one of the more memorable winters of the last 50 years, when the whole country suffered freezing conditions for a full three months from Boxing Day onwards. And it was two years almost to the day following the loss of the tennis courts

A further benefit to be drawn from learning first-hand about incidents of fire is the incidental narration of details unavailable from any third-person account. A conversation with Jamie started our conversation by finding out how the Cromer Brigade were alerted to the blaze.

Jamie Edghill *That afternoon, we'd been attending a chimney fire in a house next to the village school in Gresham and were on our way back to the station in Cromer. We'd got as far as the Lion's Mouth at Aylmerton, when a message came through from HQ asking us to proceed to a fire at the Newhaven Court Hotel on the Norwich Road. As we drove over the railway bridge heading down into Cromer, we could see a low cloud of smoke hanging and drifting over the town; we knew then that we had a big job on our hands. As we came through the town we could hear the siren whirring away, summoning the second crew.*

Late afternoon in winter, it was probably getting dark and the temperatures were plummeting.

Jamie Edghill *Yes they were. When we arrived we realised it was freezing very, very sharp indeed; I was given the job of linking up with the water mains, so I took the stand-pipe, key and bar down to the hydrant on the*

corner of Cliff Avenue and Norwich Road. I tried my hardest to get the lid off, but it was frozen absolutely solid, so I had to run back to the appliance and get a sledgehammer. In the end, I had to break the cover to the hydrant and eventually I managed to connect up with the hose which was relayed up to the back of the pump.

It was that bad! How advanced was the fire at that stage?

Jamie Edghill *By then the fire was raging through the roof and burning very fiercely. Once we had secured the water supply, I was taken off the hydrant and ordered to go inside the hotel with Geoffrey Newland (known to us as Happy) and Dennis Green. Our task was to take a length of hose up to the second floor and fight the fire from there.*

Right in the thick of it.

Jamie Edghill *Yes, we were. Before long, we had quite a decent jet coming through, and we were working hard at the flames. We'd been there about ten minutes when we heard a lot of people shouting at us; eventually we realised we were being told to get out of the building quickly. What they knew, but we didn't, was that debris from the roof had fallen down on to the lower staircase and that was blazing away. So we had to make a quick getaway – that was the fastest we'd moved all day. So we jumped through the flames and fled outside through the door.*

That's what you call teamwork.

Jamie Edghill *Yes, in those kinds of situations you really have to trust in your mates. Once we'd dragged the hose out behind us, we had to continue fighting the fire from the outside, alongside the other firemen. By this time we'd been joined by pumps from Mundesley, Holt, Aylsham, North Walsham and Sheringham – there was a total of about 60 men fighting this blaze. Senior Officers came too; we even had a visit from the Chief Fire Officer himself, Mr W. M. Ward. Afterwards, he praised us with the words: 'The fire was never allowed to spread below the level where it began.'*

What about the fear that people may have been trapped inside?

Jamie Edghill *We knew that all the hotel personnel had been evacuated from the building; as it was January, only the owners, Mr and Mrs Donald Stevenson, had been inside. Apparently, they'd been sitting down in the lounge having tea when they were told smoke was pouring from the roof. We carried on fighting the fire until about nine the following morning, when things started to quieten down a little bit, but it wasn't until ten that we could relax. At this point we left one crew on the spot for damping down purposes while others were allowed to go home for a rest; later in the day we exchanged crews.*

Dealing with a car fire at Bridge Farm, Northrepps, in the 1960s. Peter Wickers is on the left, with Noel Fisher, Jamie Edghill and Les Folds.

What other problems did the extreme temperatures cause you?

Jamie Edghill *Some things from that night stand out; I hadn't been in the Brigade very long – a year, perhaps; it was freezing very sharply, so hard, in fact that the spray coming back on us froze our tunics solid, so you couldn't bend the fabric at all. That's how we had to fight that fire throughout the night. Another problem were the icy conditions underfoot – it was often difficult for us to keep our footing.*

Any idea as to the cause?

Jamie Edghill *A fire investigation team came in the following morning; they believe the fire originated in a cupboard on the first floor near the bathroom where it was filled with blankets and quilts, and the blaze had spread up into the roof via a pipe duct. But this area was so badly damaged it was difficult to be absolutely sure. The hotel was so severely damaged that it was eventually pulled down and the land redeveloped.*

Brigade Competitions

The tradition of competing in county championships continued into the early sixties. Here's a press report from Saturday 6th July 1963:

Cromer firemen romp to county contest win

A Cromer Fire Brigade team had judges checking their stopwatches with amazement at the annual competitions of the Norfolk Fire Service on Saturday. They could just not believe the time of 2 mins 52.2 secs that it took Cromer to win the pump competition. It was the first time a team had taken under three minutes. And it was only the second year that Cromer had entered. The runners-up were Hunstanton, nineteen seconds slower.

Winning team: J. Pope, A. Kirby, T. Barnes, R. Norton, L. Folds.

Jamie Edghill removes charred material from Rust's storeroom.

Rust's SATURDAY 20TH JUNE 1964

This fire occurred in the height of summer 1964, right in the centre of town. It was in Rust's drapery storeroom in Garden Street, Cromer. Early awareness of the impending disaster came in an unusual way.

*While a fire was blazing in the ground floor drapery storeroom, six employees were having lunch in the canteen above, and the first they knew of the blaze was when Mr George Palmer of North Walsham felt the floor and found it 'red hot'. (*Eastern Daily Press*)*

Staff did what they could, but were frequently forced back by the intensity of the heat. Indeed, Miss Evelyn Pearman, who worked in the store as sales assistant, was overcome by smoke as she, along with her workmates, made valiant attempts to salvage stock.

Firemen attending the Rust's fire: Station Officer H. Mitchell, A. Royall, P. Gray, K. Fairhead, H. Pearce, G. Newland, L. Folds, J. Pope, P. Wickers, T. Barnes, D. Chapman, A. Kirk, J. Bacon, D. Green, E. Thetford, C. Kemp, J. Edghill, R. Norton.

The Albany Hotel MONDAY 31ST MARCH 1969

Cromer lost one of its major landmark buildings to fire late in the 1960s. The four-star Albany Hotel, formerly the Grand Hotel, stood on the town side corner of Runton Road and Beach Road, site of the present Albany Court flats. It had 60 rooms and was the property of Mr George Knapton. The fire started in the early hours.

Jamie Edghill *As I made my way down to the Fire Station from the Metton Road, I could see a great glow in the sky. And the message was: the Albany's on fire! Two machines from Cromer were sent to the sea front straight away.*

W. Mason (referee) shakes hands with Tommy Steele in front of Chief Fire Officer R. Pearson.

Charity football

Jamie Edghill got a taste for fundraising on a grand scale with the bold plan to invite a Show Biz football team to Cromer; so, early in January 1967, he made contact with their organiser, Bill Parry, with a view to coming to Norfolk to play a charity football match at Cabbell Park in aid of the Benevolent Fund on Sunday 30th July.

Jamie Edghill *It turned out to be a brilliant day; when agreeing to the fixture, Bill Parry, their organiser, made it clear that they would not be able to guarantee any particular star to take part, but they gave a list from which they would make up the team on the day. He also reminded us that we would be playing against some valuable legs, hinting that we should go a little bit easy with our tackles! I'd organised a local girl, Penny Jewkes, to start the game for us; she'd been a pupil at Cromer High School, before making her name as a singer with the Black and White Minstrels, a variety troupe which had a regular show on TV at that time.*

Tell us about the day itself.

Jamie Edghill *The Sunday of the match was a glorious day, with a crowd of about four thousand to cheer us on; the team coach, which had brought the players up from London, drew up at Cabbell Park, and Bill came down to greet me. 'I've got a star surprise for you today!' He then announced that very shortly the one and only Tommy Steele would be stepping off the coach. This really was a coup! Tommy Steele was one of the biggest stars of those days, as singer ('Little White Bull'), stage performer in the hit musical* Half a Sixpence *and as a film actor, too. He was at the peak of his fame. He'd flown in from America the previous day, contacted Bill Parry, asking for his name to go on the team sheet. And there he was, stepping off the coach in jeans and a T-shirt, I couldn't get over it! But we shouldn't forget the others; there were four members of the group 'Freddy and the Dreamers', Tony Dalli who was appearing at the Windmill Theatre, Great Yarmouth, not forgetting Jeremy Bullock who was in* The Newcomers, *a current TV programme.*

Let's now turn to the pages of the *Eastern Daily Press* for their sizzling report of the match.

Firemen Set Out To Extinguish Showmen

Twelve shared goals was the fitting result to the Charity Football Match played yesterday. The firemen set out as though they really meant to extinguish the showmen and ran up a 3–1 lead by the interval. The prettiest piece of footwork all afternoon was when Penny Jewkes of the Black and White Minstrel Show kicked off. From then it was every man for himself.

Reg Hindry of Aysham put two goals past Derek Quinn of the Dreamers, who wore dark glasses against the glare of the sun in the Show Biz goal. Jeremy Bullock – better known as Philip Cooper of the 'Newcomers' – headed one in retaliation, and then Kenny Walker, another member of the Aylsham Brigade on target, made it 3–1.

By this time it looked as though the only chance the stars had was for the siren to call the opposition away. But what a revival in the second half. Tommy Steele turned on half a sixpence to pull one back and Dave Bates – another of the four Dreamers on the field – made it 3–3.

But three goals in as many minutes – two more from Reg Hindry and the sixth from Peter Punchard of Acle – turned the tables again. All was not lost, however; in true style, the Show Biz boys summoned their flagging spirits, stirred their tired limbs and the game went on to be drawn. Tommy Steele inspired the side with two more goals – the second of which was scored with Jeremy Bullock blatantly offside stroking a pigeon which had wandered on to the field.

Aided by Andrew Ray, they sought the equaliser which came from music publisher Franklin Boyd. He slammed the ball past the firemen's keeper, Jamie Edghill. The crowd of nearly four thousand were satisfied with the result, which was not 27–3, as stated by Bill Parry, the Show Biz trainer who doubled as commentator.

The only snag in the planning came when they found out that it was illegal, at that time, to levy an entrance charge for an event like this on a Sunday. They had to solve that one by selling programmes at the gate instead! Jamie went on to organise two similar events at Yarmouth and Gorleston, where the attendances were nearer 8,000.

Penny Jewkes kicks off.

BELOW: *Kenny Fairhead, John Bullen and William Cox survey damage to the top floor of the Albany Hotel.*

Once we arrived at the fire scene they decided immediately to make up a number of pumps, because of the size of the hotel.

Everybody's first thought must be the possibility of somebody trapped inside.

Jamie Edghill *You're right. We were immediately told that Mr Martin Lax, the caretaker, was inside the hotel asleep; a further problem was that he was partially deaf. One of our men was soon up a ladder to the rear of the hotel, trying to locate which room he might be in; however, close neighbours John Bridges and wife Mickey (daughter of Mr Knapton) had already anticipated the problem, and were quickly on the scene; as the fire was burning at the very top, they were able to enter the hotel, go straight to his room and wake him up. With the help of the firemen they could lead him to safety.*

A huge relief all round.

Jamie Edghill *After this early success it was our job to get hard to work and extinguish the fire itself. But we immediately faced another problem, when* *it was discovered that a gas feeder pipe was fractured – we didn't want an explosion too.*

Some quick thinking, then.

Jamie Edghill *Yes, Les Folds forced his way down to the basement through thick smoke to turn off main gas supply – and none too soon.*

I understand there were over 70 firemen tackling the blaze at its height.

Jamie Edghill *There probably were. Other brigades arrived in quick succession – appliances from Holt, Sheringham, Aylsham, North Walsham, Mundesley; I think Acle were present, too, probably with breathing apparatus and replacement cylinders.*

You say the fire was at its fiercest in the upper storeys.

Jamie Edghill *Yes, we discovered that the fire had started somewhere in the lift shaft; but because of the upward draught there, it had gone straight up to the roof and spread all the way along. We had to tackle the blaze from the*

Firemen attending the Albany Hotel fire: Kenny Fairhead, Joe Pope, Tom Barnes, Dennis Green, Peter Wickers, Jamie Edghill, Les Folds, John Bullen, William Cox, Pat Taylor, Donald Chapman, Keith Barker, Richard Davies, Geoffrey Newland, Alfred Kirk, Jeff Morris.

outside with long extension ladders, and from inside as well. I've actually got the press report here.

Go on.

Jamie Edghill *The fact that many of the bedroom doors were closed had helped to prevent the spread of the blaze. A wave of heat went along the corridors, blistering walls and doors, and causing fittings to crumble. Some rooms, their doors buckled by the heat, were comparatively little damaged. Firemen courageously made their way to the top floor inferno to master the blaze; others directed jets through windows from 45 ft ladders. The fire was finally beaten as firemen forced their way through intense heat at the top and using powerful jets of water, stopped the flames from running along the corridors.*

That gives a very graphic description. It must have gone on for several hours.

Jamie Edghill *It did. When daylight came, it was clear that the fire had now been extinguished; nevertheless we came across smouldering parts – hotspots – which had to be dealt with. And there are always other potential dangers to be aware of: chimneys left standing on their own, water tanks in the roof space which might topple. We also did the usual job of making sure the fire didn't break out anywhere else. We were able to get a break at about ten in the morning, but then had to return to continue checking the state of the hotel.*

When you look back, what's your overall impression?

Jamie Edghill *The destruction of the Albany in 1969 was my first really big fire; it represented a major loss to the town. It was the biggest fire I had attended; I was already seven years into my service, but I found it quite frightening: the flames were going through the roof and rising to thirty feet in the air.*

Dutch exchange visits

There became a growing awareness in the 1970s of the benefits to be had from reciprocal arrangements with communities in other countries. This took the form of 'twinning' between towns, or links that were profession-based. Cromer Fire Brigade joined an initiative set up by Aylsham for exchange visits between football teams from The Hague and Norfolk, with accommodation in the homes of their hosts.

Jamie Edghill *We took the ferry to the Hook of Holland, arriving quite early the following morning. Our schedule included three matches in two days – a tough schedule! We were given a warm welcome at the Hague Fire Brigade, then we went straight into the first match against their team, who unfortu-*

October 1971.

nately won 3–0. Later in the afternoon we played a Rotterdam XI, who beat us 2–1. The following morning's match we managed a draw 0–0. So at least you could say we were improving!

They gave us a wonderful time off the pitch; of the many interesting places we visited the most memorable was the Madurodam, a model village quite unlike anything we have in this country – more like a model city! It covers a huge area, and includes an airport with planes made to scale, moving vehicles, fire tenders, everything. A major attraction proved to be a large ship floating in a waterway; halfway down the deck was a hatch that opened, and lo and behold, flames shot into the sky! But never fear, a fireboat was summoned, pulled up alongside, and proceeded to spray off four jets of water; within about five minutes the fire was extinguished! Mission accomplished, the fireboat then moved off downstream. Absolutely amazing.

Our party consisted of: Kenny Walker, Colin Spink and Reggie Hindry from Aylsham; William Cox, John Bullen and myself from Cromer, Pat Pipe from North Walsham, Pat Penn from Wroxham, two men from Acle, Barry Marsh and Jack Jarvis, then Terry Tubby from Martham. The whole trip was very well organised, with a great atmosphere, and we certainly had a wonderful time. So much so that the following year we invited them back to England. In all, there were three or four return visits.

The ship in the Madurodam model village.

Opening of the new fire station

The new station was opened on 9th October 1971. It was designed by G. C. Haydon ARIBA, County Architect, to the requirements of the Norfolk Fire Service. It was opened by Mr W. J. Fulcher, a former Cromer JP, councillor and member of the County Fire Service Committee for several years. After he'd cut the traditional tape, the fire siren sounded and the tenders turned out for a fire drill. Mr Fulcher headed a list of guests including the Chief Fire Officer for Norfolk, Mr R. L. Pearson, the chairman of the Fire Service Committee, Mr E. W. Bartram, and the County Council chairman, Mr W. J. Hayden, and Clerk, Mr R. A. Beckett.

Town Mayor Norman Louch and his wife speak to Divisional Officer Don Wallace of Norfolk Fire Service; Mr Fulcher prepared to cut the ribbon in the centre as Chief Fire Officer Bob Pearson watches on his left.

Fundraising for the Firemen's Benevolent Fund

As I suggested earlier, any fundraising activity undertaken on behalf of the Fire Service had other aims and produced other effects – notably in the area of enhancing community awareness. It is very clear from the accounts of these very different activities that Cromer people came to understand and appreciate the work of the Fire Service to a much greater degree. We will now take a glimpse back at some of these activities, with the realisation that they were not one-offs, but happened on a regular basis.

Fire station Open Days

For many years in succession, the new fire station was opened to the public to enable them to view the town's Fire Brigade at close quarters. One year, the Firehouse Five – an ex-fire service country group from Mansfield, Nottinghamshire – came along to provide musical entertainment, while children climbed all over the appliances on display. They were visited, too, by Welephant – 'the elephant who never forgets fire safety'. Also represented were the other emergency services: police, ambulance and coastguard, while RAF Coltishall also brought a fire engine for viewing. On this occasion the Seaside Special stars Gordon and Bunny Jay opened the proceedings. Typically, the event would raise £1,200 for the Benevolent Fund.

ABOVE: *Open Day: Gordon Bowles, Graham Lee, Dave Roberts and Mark Smith.*

Wine and Fish evenings at the Cliftonville Hotel

A regular event that became very popular was the Wine and Fish evening in the upstairs of the Cliftonville Hotel. Mr Joe Cole of Lowestoft would bring a wide range of fish, already cooked and prepared, to which people would help themselves. The addition of a glass of wine made it a very good evening. He would end up with an auction of raw fish, displayed on a big slab; people had great fun bidding and they could pick up some really good bargains; probably a hundred people would attend each session.

Guy Fawkes Night

In the early eighties Jamie became involved with Suffield Park Infants School as a parent, and it was a natural move for him to take responsibility for Guy Fawkes night – to ensure it was a safe evening for everybody. They were big events; an average attendance would be 600 to 800 people, all gathering in the field behind the school for an enjoyable evening. Associated with that was a Poster Competition for each of the classes in the school.

This particular arrangement lasted for nearly twenty years. Moreover, the tradition continues, but in a different location, the football ground.

Meadow Playgroup

For several years running in the mid-1970s, Jamie invited members of the Meadow playgroup, which met in the Methodist Hall, to visit the Fire Station.

A visit from the Meadow Playgroup in the 1970s, with Jamie Edghill on the left and daughter Michelle standing in front of him.

Jamie Edghill *Another rewarding activity for me was to invite children of between three and five years old from the Meadow Playgroup to visit the station. I showed them all around the station, but of course they all particularly enjoyed sitting behind the steering-wheel and pretend they were driving the machine. I even had them put on firefighting gear; you could hardly see them sometimes, lost inside these huge tunics – they almost covered them completely! Another activity that appealed to them greatly was to get the first aid reel out, run it with water and give them little targets to knock down. We'd round off the morning with a few refreshments – orange squash and biscuits. Mothers and playgroup helpers got as much out of their time as the children, I think – there was always plenty to see and do. I used to explain as much as I felt was within the scope of young children to understand.*

Endless scope for fun!

Jamie Edghill *We were no longer in the era of bells, but the appliances were fitted with two-tone horns which the children were allowed to operate. That used to alarm them a little at first, but once they got used to the position of the button placed on the floor near the driver's foot-pedals, there was no stopping them! They'd kneel down and press it furiously with their hands; I often had to quickly go and turn it off, the noise was so loud for everybody else standing nearby.*

Did you ever get a call-out during one of these visits?

Jamie Edghill *No, we didn't! The ideal timing would have been towards the end, when they would have seen us take a tally from the board, get into our uniforms and get away from the station as quickly as possible. I have to say it would be a great deal more difficult to arrange such a visit these days, with all the Health and Safety issues which are now central to the planning of any outings.*

Fundraising statuettes

A very personal approach to fundraising was undertaken by Jamie over a number of years, proving to be both very effective and popular with Fire Services at home and abroad. He found a way of creating statuettes – different models of firemen – from a resin mix, then adding colour to the resulting figure. Here he gives the background to this endeavour.

Jamie Edghill *At the end of our visit to the Dutch Fire Service we were giv-*

en a statue of a Dutch fireman, and I started to wonder whether there might be scope to create something similar of an English one. I made enquiries both locally and nationally, but drew blank. After about a year I heard about a gentleman called John Pooler who might be able to help me. So I went to see him at his home, taking my Dutch statue and photos with me. I soon discovered that he was an ex-fireman from Nottingham, and he was prepared to make a model from which I would be able to take a cast. So that's what he did.

You hit lucky then!

Jamie Edghill *I certainly did. A few weeks later I went to see the figure which he'd produced – it was absolutely perfect. From this he was able to make a mould, from which I would be able to produce replicas. The whole idea was to sell them with the proceeds going to the Benevolent Fund.*

Tell me about the process.

Jamie Edghill *Each statue took me about eight hours to produce. I would take the resin mixture, add the catalyst and the brass, bronze or nickel powder, heat it all up for it to go hard. Once out of the mould, they would need to be tidied up, stained with a dark stain and then buffed up till shiny on a machine. Then they were ready for sale to firemen across the country.*

Who were your customers?

Jamie Edghill *Chief Fire Officers mainly! One special destination was the House of Lords – I had a phone call from a top man with that commission! They went on to prove very popular as gifts when Chief Fire Officers went on exchange visits abroad. So I know my statues are all over the world – Germany, Africa, New Zealand, Australia, Canada, America – and that's not all.*

A job for life!

Jamie Edghill *I carried on doing this for several years, but there were snags in the production process, mainly the fumes given off by the mixture. My 'factory' was a small shed in the garden with the door left open! John Pooler went on to make four different statues altogether; the first one of an ordinary English fireman, the second wearing breathing apparatus, another one was of a fireman kneeling down supporting a young child rescued from a fire, and the fourth was a fireman holding a branch. People always said how pleased they were with them.*

How many did you make altogether?

Jamie Edghill *I made about five hundred statues in all, probably. The cost of materials was high and always seemed to be increasing, so I was forced to raise the price regularly over the years I made them.*

ABOVE: *Jamie with the Margaret Florence Waters trophy, awarded to him in January 1990 for services to the Benevolent Fund. His fundraising efforts netted more than £20,000 for the fund over two decades.*

Incidents of the 1970s

Royal Links Pavilion WEDNESDAY 5TH APRIL 1978

The Royal Links Pavilion along the Overstrand Road (on the site of the present Country Club) was built in 1926 as the ballroom to the nearby Royal Links Hotel. When that burnt down, the ballroom became a nightspot, often entertaining famous pop groups on tour. There was a considerable rumpus at the prospect of a Sex Pistols concert, but the group promised to be on their best behaviour and the event went ahead without trouble. But on 6th April 1978 people read the press headline 'Smoke Dims Town as 30 Fight Fire'.

Alfred 'Leicester' Kirk wears breathing apparaus and William Cox carries its control board (recording entry and exit times) at a fire at the East Runton Washeteria in the 1970s.

Jamie Edghill *There were many factors that made this a very strange incident. The whole thing started at about 3 o'clock – I remember it well, I was working in Cromer at the time. We had a call to a grass fire at Sidestrand, so on our way we passed the Royal Links on our left. We spotted a tractor and trailer backed up to the main entrance and could see men loading up chairs and tables; we all commented on this, it looked as though they were moving out. But we didn't think any more of it; we extinguished the grass fire, we made our way back, we passed the Royal Links again; everything seemed to be quiet, there was nobody about. I went back to work, and was just about to leave off at about 5.30, when the alarm went off. The first one in took the call, only to learn that the Royal Links Pavilion was alight. We remembered those men.*

You thought there might be a connection?

Jamie Edghill *Well, we took both appliances and when we arrived we found flames coming up through the roof; as it was timber the fire was going quite well. However, another thing we noticed was that the fire was not confined to one area; the whole length of the building was ablaze. That immediately gave us suspicions about the origin of the fire.*

The fire had really taken hold, I suppose.

Connecting the featherweight pump to the hydrant at Crossdale Street, Northrepps, in the 1970s.

Jamie Edghill *Yes it had. By now we had four pumps in attendance, including Sheringham and Mundesley, and this gave us a total of about 30 firefighters. We were able to run out hose and direct jets on to the flames to extinguish everything we could; and within an hour we had the blaze under control.*

And what about your suspicions?

Jamie Edghill *We had to get the Fire Investigation Team in; this was a very historical building and important to the town; actually, it was built to go with the hotel which had been burnt to the ground many years ago. The building was totally destroyed and eventually demolished. At the time it was up for sale. It has to be said, everything pointed to arson, but I don't know if anybody was charged.*

Overstrand Court Hotel FRIDAY 13TH OCTOBER 1978

Sometimes the call-out comes at just the wrong moment! The Fire Brigade's darts team were enjoying a competition at the Colne House Hotel between teams drawn from the local emergency services – police, ambulance, fire service, lifeboat, RAF Coltishall, when they got the call at about 10.30 p.m.

A darts match interrupted – from the left, William Cox, Donnie Chapman, Jamie Edghill, Richard Davies and Billy Davies (the last two later became coxswains of Cromer lifeboat).

Jamie Edghill *When we arrived in Overstrand we knew we had quite a major fire on our hands, with the top floor all well alight. Obviously our first job was to see that everybody was out of the hotel, because we had been told that some of the people were asleep when the fire broke out. A lot of people had been in the bar, and once we were satisfied that everyone was out, we could start tackling the fire. It certainly was severe, so we had to make up other stations to help us out. In all we had the two appliances from Cromer, plus Sheringham, Holt, Aylsham, Wroxham, North Walsham – altogether perhaps 50 firemen attending this incident.*

Was it the smell of smoke that alerted people of the fire in the roof?

Jamie Edghill *The owner of the hotel, Mr Northway, had not realised that he had a fire on his hands immediately, but he knew he had a problem of some kind: the hotel lights had gone out at about ten o'clock, he'd had to provide some emergency lighting and he'd also ordered an electrician and the Eastern Electricity Board. Little did he know at that stage that there was a blaze up in the roof void. But once the alarms were sounded about 30 minutes later he realised the true nature of his problem. We remained on duty until the early hours of the morning, when one crew stayed behind while the other one had a break; then we returned to relieve them later on. The building itself was not a complete wreck; we did quite a decent job in extinguishing the roof, which could be re-built.*

Boy who got his head stuck

And now for a very unfortunate thing to happen on holiday.

Jamie Edghill *This incident involved a young lad who got his head trapped in the turnstiles of Cromer Pier. We actually received the call while we were drilling in the yard at the old Fire Station in Canada Road. When we got down to the promenade, we found this poor boy – of about eight or nine years old – bent over with his head caught in the turnstile. It looked quite comical, but it certainly wasn't funny for him! We talked to him all the while, and there was not a tear to be seen. Anyway, with the help of our hydraulic equipment we were able to lever up from the concrete slab, so that the metal gave; we could then bring his head through the widest section. As soon as he was clear, he turned and just ran to his mother, when he just burst into tears – there were floods in all directions! Quite a crowd had gathered, you may be sure, everybody was intrigued as to how he might be released. Not a scratch on him!*

Incidents of the 1980s and after

Cromer fire service in 1982

BACK ROW, *left to right: Richard West, John Bullock, Dave Roberts, Mark Smith, John Bullen, Alan Bumphrey, John Cooper, Neil Babbage, Donald Chapman, David Abbs, Gordon Bowles.*

FRONT ROW: *John Balls, Tom Barnes, Jamie Edghill, Peter Wickers, Jeff Morris, William Cox.*

Norfolk Textured Yarns TUESDAY 14TH AUGUST 1984

A major blaze destroyed part of the premises of Norfolk Textured Yarns, a factory off the Holt Road which had opened in the early 1960s, belonging to the brothers Stuart and Graham Johnson. The business usually worked 24-hour shifts, but as it was the holiday season, it was on day-time operation only.

The alert came late in the evening.

Jamie Edghill *It was an evening call-out, at around 10.30 p.m. I was at home in Howards Hill Close, and for about ten minutes I'd been hearing a noise very much like shotguns going off; I'd even stepped outside the back door to see what it was all about, without reaching any conclusion. Then, back indoors, my pocket alerter went off, and on my way down to the station I reckoned there must be a connection between those noises and the call – without being able to figure it out exactly at that stage. Once there, I discovered that the location for the fire was the textile factory on the Holt Road; and actually that lies just the other side of Howards Hill. Once we arrived, we could see immediately that this was a big fire, so we made up for other pumps. As for the sounds of gunfire, we now knew that they were the explosions caused by the asbestos roof in the heat.*

A case for special equipment?

Jamie Edghill *Yes, the emergency rescue tender came from Norwich and the snorkel was summoned, too. They were in addition to the appliances backing us up from Sheringham, Aylsham, Holt, Mundesley, North Walsham, Fakenham, even as far as Wells. In all there were about 70 firefighters and eleven appliances. Let me just explain the term 'snorkel'. The firemen get into a cage when it's at a low level, it's raised by a crane, allowing us to tackle the blaze from above.*

Let me just read from the press report. It says: 'As flames threatened to spread to another part of the building, about twenty firemen wearing breathing equipment moved in to the southern end to try to prevent the fire advancing.' Can you elaborate on this?

Peter Wickers, Station Officer 1975–82.

Firemen attending the fire at Norfolk Textured Yarns: G. Morris, G. Lee, D. Abbs, M. Smith, J. Balls, J. Bullen, D. Roberts, J. Edghill, G. Bowles, J. Cooper, P. Abbs, W. Cox, S. Morris, D. Chapman.

Jamie Edghill *I can – I actually played a role in this – I was operating the breathing apparatus board. This means I had to time the firemen in, and keep a check on the time they should be out of the building; with each breathing set there is a DSU – a Distress Signal Unit which sounds when there's ten minutes left on the cylinder. That's all well and good, but if it's taken a man 15 minutes to get to his current position, ten minutes aren't enough to get out. And there was another problem with this incident – there was machinery all over the factory floor. So, as a further precaution, I had four firemen in breathing equipment standing next to me, and if we had any worries we would send in two in order to check that the men hadn't become trapped in any way. At one point we got very close indeed to the limit, and I had two men just fitting their masks prior to going inside, when those inside appeared at the doorway.*

Was the building totally destroyed?

Jamie Edghill *Part of it was. The ferocious blaze at the northern end was allowed to burn itself out while we concentrated on saving the rest of the building. The fire blazed until the early hours of the morning, so it was a long time before any of the appliances were allowed to leave the scene.*

Did anybody discover the cause?

Jamie Edghill *We did find a window had been left open, so that aroused some suspicion; an intruder perhaps, but nothing conclusive.*

Jewson's 17TH FEBRUARY 1987
William Cox contributed to our researches, too!

William Cox *We received the call in the early hours of the morning, and by the time we arrived there was a very good fire going; it had broken out about ten yards into the building, on the left. We realised quite soon that the outbreak had started in the paint store – which accounted for its strength. The blaze destroyed much of the office space, along with the paperwork. Although we thought it unlikely that there was anybody in the building, you can never be totally sure; in the early stages of fighting the fire you have to have all possibilities in your mind.*

What stands out in your mind?

William Cox *There was no fire at all outside, but inside there were little blazes at intervals; once we'd got the whole situation under control all the indications were that it was arson; we found several incendiary devices which had not ignited, both inside and outside. These devices were quite straightforward – matchboxes with several matches sticking out, that's all, but evidently very successful. For example, there were pallets lying about – some with building materials, some empty, but we found devices here, too.*

William Cox joined Cromer Fire Service in 1966, having already had four years with the Mundesley Brigade, making it a full 40 years in the service. For 20 of those years, William was also a member of the lifeboat crew. Station Officer from 1985 to 2002, he was awarded the MBE in 2002.

It all sounds very deliberate.

William Cox *Thoughts went immediately to whoever might be the culprit – an employee with a grievance, perhaps, or a customer who owed a large amount of money. It was certainly somebody with a knowledge of the internal layout of the premises. There was tremendous damage to the stock, both from the fire itself and from the water. A large number of pumps attended this incident – eight or nine, I believe. A whole-time officer arrived on the scene, and once the blaze was extinguished we looked around the building together.*

Church Street flats 6TH AUGUST 1987
Cromer Brigade received a Certificate of Commendation for their handling of this incident, in which they rescued 13 people.

William Cox *On arrival we saw the fire had taken hold just behind the two big double doors; we forced these open and found the entrance hall full of dense smoke; it was a large space with a staircase at the far end that split in two; we believe the property had been built as a hotel. We fought the fire here with hose reels and breathing apparatus, while another crew was sent round to the back of the building to investigate. By the time they returned, our second pump had arrived; they were ordered to take their ladders round to the back for possible rescue work – there was certainly no way out down that front staircase. But access round the back was actually quite a distance – up Mount Street, left into the lane that runs behind the buildings, that way.*

You really need to be familiar with the buildings.

William Cox *Yes, you do. There was a flat roof to the rear of this block, so our crew had to escort the people trapped upstairs through the windows, on to the roof, then down the ladders. Of course, nobody was absolutely sure who was in the building at the time – they couldn't be, it was divided into flats; so-and-so may well have been in or out, such and such a flat may have had visitors, whatever. So as soon as the fire had been contained downstairs, lads in breathing apparatus searched the place until we had verified that there was nobody else up there.*

Do you remember the people now?

William Cox *I remember one particular lady whom we rescued; she was quite a well-known character in the town, rather a large lady. She did make it through the window on to the roof, but it was a bit of a tight squeeze – but she took it in good part.*

Keeping everybody calm all the time.

William Cox *Getting people down the ladders can sometimes be a drama in itself; you have to place a firefighter a couple of rungs below and keep talking, keep reassuring.*

Firemen attending the incident at Church Street:
W. Cox, J. Baker, G. Lee, D. Abbs, M. Smith, P. Green, D. Chapman, N. Babbage, P. Abbs, J. Cooper, J. Balls, D. Roberts, G. Crisp.

And the cause of this incident?

William Cox *Later we discovered that the fire had started – probably late at night – in rubbish left in the hallway just inside those double doors. There had been a pile of black sacks, a couple of old mattresses, a carpet, a few bikes, that sort of thing, all of which somehow had been ignited – through negligence, stupidity, a cigarette end – we couldn't be sure.*

Chip pan demonstration CROMER CARNIVAL 1988

The Fire Service involvement with Cromer Carnival goes back many years – sometimes in the form of a float in the procession, perhaps as entrants in races. In 1988, Jamie gave a demonstration of a chip-pan fire – to spectacular effect!

See colour picture on front cover.

Jamie Edghill *This followed a visit I had made while on holiday in Mansfield the previous year. On that occasion I saw a similar demonstration which showed what happens when water is thrown by mistake on a chip pan already well alight. On my return to Cromer I set about planning for the following year's carnival. I had a special unit made up which included a recessed pan for the fat. I stood in the middle of the arena and told the crowd what I was hoping to show. This was a key element in the programme of festivities held on the Runton Road car park. The moment came to set the pan alight – just think if this happened in your kitchen!*

The cottage on Overstrand cliffs 11TH JANUARY 1990

You never know what to expect!

Jeffrey Morris, a former policeman, was Station Officer from January 1983 and also served on the inshore lifeboat crew. He died at the age of 48, just two years after becoming Station Officer.

Jamie Edghill *At about seven o'clock we received a call to a house on the old Coast Road at Overstrand. We were responding to what was described as a kitchen fire. When we started tackling it we knew that it wasn't too serious an incident – there was smoke drifting around throughout the interior, but we were able to extinguish the flames quite quickly. Then we went through the downstairs, ventilating the rooms by opening the windows. Two colleagues and I then went through to the room at the back, which had a view out to sea. Something made us look down at the far end of the room, and we could see daylight; taken aback, we went over to it, bent down and had a closer look. What we were in fact looking at through a hole in the floor was the sea a hundred feet below! We stepped back – and then we each realised we were probably standing in a house whose foundations had disappeared down the cliff!*

A considerable shock.

Jamie Edghill *Never had we retraced our steps so carefully! We rejoined the others, telling them to avoid going into the back room at all costs! Clearly, we wanted to see exactly what the situation was, so we went round the outside* *to study the rear of the building. Yes, six feet of building overhanging the cliff – there was nothing underneath it at all.*

A bit more than a kitchen fire!

Jamie Edghill *Obviously, the gentleman in residence had to be persuaded to move out – something he did not like in the least. However, it was not many weeks before we heard that his house had gone altogether. I can't recall the man's name – just his nickname – Tin Hat. That's because he always wore a safety helmet: he knew more than he let on!*

Gas canister explosion 11TH APRIL 1993
Some incidents leave the hair standing on end.

Jamie Edghill *We arrived at the station to hear that a gas explosion had taken place at Warren Court, just off the Overstrand Road in Cromer. I went out in charge of the first appliance, and we were quickly followed by the second, with Station Officer Cox on board. It was about 4.30 in the afternoon. As in most cases, we were not sure what we might find there on arrival, but to my amazement, as we turned into the Warren what we encountered was quite simply unbelievable. Apparently the owner had been using a gas cylinder in an upstairs room; this had led to a build-up of gas, setting up an explosion which had ripped the whole of the brick gable end out. I had never seen anything like it – it was incredible what a small gas canister could do.*

The photos say it all.

Jamie Edghill *Our first dilemma was to work out what our priority should be. Immediately, two of the crew went up to the room to see who might be there, and to check that nobody was injured. The next essential was to tackle the rubble; by that I mean we had to move it brick by brick just in case, at the time of the explosion, there had been a passer-by who might still be lying underneath the heaps of brickwork.*

Did you find anybody?

Jamie Edghill *Well, the homeowner or tenant – the other members of the crew were attending to him; apparently, he'd been working for most of the day in another room. He had some minor burns, which were treated in hospital, that's all – he was very lucky indeed. The only other person involved was a lady, thought to be his mother, not injured as such, but in considerable shock. Our conclusion was that it was probably a faulty canister – a leak in a connection, something like that.*

This was the last major incident Jamie was involved in before his retirement.

Cromer Fire Brigade in 1993:

BACK ROW: *S. Morris, M. Smith, R. Chapman, P. Green, H. Khalil, S. Hedge, G. Lee, D. Cooper, J. Baker.*

MIDDLE ROW: *D. Abbs, J. Edghill, P. Abbs, N. Babbage, W. Cox, G. Crisp, J. Cooper, G. Bowles.*

FRONT ROW: *D. Roberts, D. Fisher.*

Edinburgh House flats SATURDAY 29TH JULY 2000

This was a major fire in a very public part of town.

Willy Cox *When we got there, we pulled up right outside on the corner, just before the Red Lion; it's very narrow there; if we'd have known what we were in for, we wouldn't have chosen that spot! I'd thought initially that it was a cooker, or a frying-pan. We could see a glow though the kitchen window; two of the lads had already got their BA (breathing apparatus) sets on, and they went down seven or eight steps to the kitchen door into the property. I stood at the top of the stairs and watched as they took a hose down with them; when they'd got the door open, they were amazed to discover the seriousness of the blaze – there was no way they could enter, it was far too bad. The helmet visor of one of the lads had buckled with the heat! And that was just from standing at the doorway! We got an inch and three-quarter hose rigged up, made up for a second pump.*

A lot to take in all at once.

William Cox *There certainly was – the adjoining property, for one. We knew the fire was spreading up though the ceiling of the basement room where it had started from the smoke appearing here. We'd been speaking with some of the occupants who had come out, and we learnt that there might be somebody in one of the flats above, as they'd heard footsteps up there the previous evening. So we had to follow the rule-book and notify headquarters of the possibility of persons being about.*

A case for extra help.

William Cox *Yes. Some of the lads had gone down the Red Lion steps to the promenade to look back up at the building – but its location on the cliff-edge made it extremely difficult to see into. It wasn't long before a whole-time officer arrived on the scene; he made up more pumps again, including a turntable with a snorkel.*

Right from the time of your arrival it had been such a fierce blaze.

William Cox *Yes, some of that was due to the fact that many of the internal walls were wood-fronted – boards on battens; this made a funnel between the brickwork and the wood; another factor was the removal of the chimney breast in the loft – so, what with one thing and another, there was an open flue up at the top of these four flats; so what had started in the kitchen of the bottom flat was soon spreading throughout the building. It was a very good job there was nobody upstairs – we wouldn't have got them out, it was just too ferocious.*

Any idea of the cause in this instance?

Neil Babbage joined the Fire Service in 1978, becoming Station Officer in 2002 – a post later restyled Station Manager.

William Cox *The fire probably started through a candle not being extinguished properly; apparently there had been a lot of these smelly candles in the room the evening before, and by the time we got the call at about one in the morning, the fire was well established. We stayed well beyond breakfast into the following day – one or other of the Cromer pumps at least, while the investigation teams did their work.*

From then till now

The 40+ years since Jamie joined in 1962 have seen major changes in the Fire Service, particularly in equipment. In our last conversation, he reviews some of the major developments.

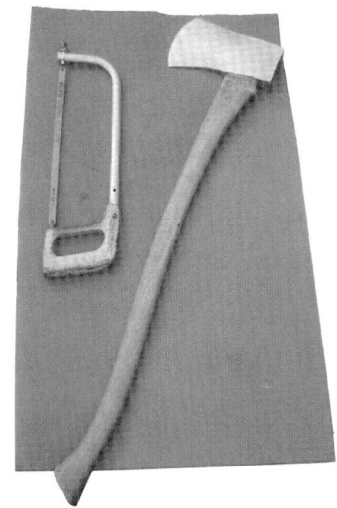

Equipment for cutting vehicles – then and now.

Jamie Edghill *One very old appliance still in use when I joined was an old pump escape with a large brass bell on the outside above the cab, operated by twisting a D-shaped handle; this was placed just above the seat taken by the Officer-in-Charge. Nowadays it's a push-button job.*

What was this old appliance like to run?

Jamie Edghill *It wouldn't! It was very temperamental; we often had to get the public to help us give it a push start down the road when we had a call-out – it was all very embarrassing.*

I imagine the equipment for dealing with road traffic accidents has changed.

Jamie Edghill *It certainly has. I can remember an occasion in the late 60s when we were called out to the Runton Road to a mini van involved in a collision. All we had to release the driver was a large axe to cut along the side of the van, and a hacksaw. It took us an hour and a half. Nowadays, each pump is equipped with Hydraulic Rescue Equipment, which would take 15 minutes.*

Call-outs must be very different, too.

Jamie Edghill *Yes. It was through the siren in the early days. The first fireman into the station would take the tally, go to the watchroom, pick up the phone with a direct line to the Control Room at Hethersett, then write down all the necessary information.*

Now it's all through computer?

Jamie Edghill *Yes, and personal alerters. Once firemen get the signal,* *they respond to the station; the first one in goes to the watchroom and takes the computer printout with all the latest information needed.*

What about personal equipment?

Jamie Edghill *Breathing sets – that's the biggest change. In 1962, only a few firemen used breathing apparatus when tackling a fire. We were expected to eat smoke!*

No laughing matter.

Jamie Edghill *No, there's much greater awareness now of poisonous fumes given off by things like garden shed chemicals, liquefied gasses and other volatile materials; and of radioactive materials, even. And boots.*

Boots?

Jamie Edghill *They've gone back to leather! They tried rubber ones with steel caps, but switched back to leather in 2000.*

And other big improvements?

Jamie Edghill *The Thermal Image Camera (T.I.C.) has made a major difference. Before, whenever we were called to an incident where there were 'persons reported', we would go in with breathing apparatus just searching and feeling with hands and feet throughout the whole building until we had found the persons involved. It wasn't easy.*

But now...

Jamie Edghill *Now each station is equipped with a T.I.C. which can pick up a person's body heat – this speeds up the search.*

What about training facilities?

Jamie Edghill *There's the county training centre at Bowthorpe, and several others for breathing apparatus training throughout the county.*

Call-outs 1st April 2005 to 31st March 2006

Domestic building fires	17
Industrial/business units	7
Road traffic collisions:	
* trapped*	9
* casualty in vehicle*	12
Animal rescues	2
Gas leak (Social Service	
* emergency)*	2
Automatic fire alarms	37
* (false alarm good intent)*	
Domestic smoke alarm	8
Vehicle fires	8
Special services	10
Chimney fires	11
Railway incidents	1
Stack/grass fires	7
Rubbish fires	4
Aircraft incidents	2
Stand-bys	1
Chemical incidents	1
Controlled burning	4

Training is also undertaken at this breathing apparatus centre at North Walsham.

Tribute to Simon Hedge

Simon Hedge, a Cromer firefighter for 14 years, died on Monday 22nd May 2006 on his way home from work as a result of a road accident. Simon was a very happy-go-lucky person who was always there to help anyone with a problem. He was Vice-chairman of the Cromer Carnival Committee and spent many hours of his own time organising and helping to bring joy and happiness to thousands of people during Carnival Week.

Simon was always ready to help with raising money for charity with the Fire Service Benevolent Fund and East Anglia Air Ambulance with his outstanding display of Christmas lights on his own house and four other terraced houses next to his in Central Road.

His abiding love was for his family; he was married for seven years to Michelle, the daughter of Jamie Edghill, and had two children, Ellen (5 at the time of her father's death) and David (3).

The funeral service took place at Cromer parish church which was filled to capacity with family, friends and colleagues. Simon's widow walked from the Fire Station with Peter Abbs from the Norfolk Fire Service leading the way to the church, followed by Simon's coffin carried on an historic turntable ladder fire engine. At the church a Guard of Honour awaited them. Pall-bearers from the local crew carried the coffin inside as townsfolk lined the streets and peered from shop doorways to watch the cortège.

For the last five years Simon had been a manager for barn conversion specialists Avada Homes; their managing director Mr Ian Johnstone paid a warm tribute saying he was a lovely character and one of life's good guys.

Simon's death was a great loss to the Fire Service; he lived for other people rather than himself.

Index

Abbs, David 84, 85, 87, 90
Abbs, Peter 85, 87, 90, 94
air raid precautions 39, 42
air raid shelters 43, 49
aircraft 56
Albany Hotel 71–6
Aldborough 30
Alexander, A. 63
Alexander, Ian 61
Alexander, Ivy 45
Allen, Hilda 50
Allen, T. 25, 33, 36
ambulance service 29
Amis, Ida 44
arson 61–2
Ashenden, Susan 51
Askew, Bill 45
Atkinson, John 53
Auxiliary Fire Service 44, 46–7, 49
Aylmerton 35

Babbage, Neil 84, 87, 90, 91
Bacon, H. 33
Bacon, J. 71
Baker, J. 87, 90
Baldro 37
Balls, Ellen 45
Balls, John 84, 85, 87
Balls, R. J. (Bob) 25, 30, 33, 36, 37, 63
Balls, W. H. 55, 57–8
Barber, Arthur 53
Barbon's Fire Office 4
Barclay, Gilbert 45
Barclay, Henry A. 12, 13–14
Barker, Keith 76
Barnes, Tom 61, 65–6, 70, 71, 76, 84
Bartram, E. W. 78
Bastow, S. 33
Beckett, R. A. 78

Beckham Palace (workhouse) 12
Beef Meadow 12
bells
    church 12, 14, 19
    curfew 4
    electric 28, 33, 36, 46, 65
    on fire engines 92
Benevolent Fund 55, 72–3, 79–81
Bessingham 39, 41
Bird, Catherine 44
Blythe, J. 25
bombs 47, 50, 53, 54, 56
boots 23, 93
Bowles, Gordon 79, 84, 85, 90
Bowthorpe 93
Braidwood, James 5
breathing apparatus 56, 86, 91, 92; *see also* smoke helmets
Bridges, John & Micky 74
buckets 38
Bullen, John 74, 76, 77, 84, 85
Bullen, R. Victor 25, 32, 33
Bullock, Jeremy 72–3
Bullock, John 84
Bumphrey, Alan 84

Cabbell, Benjamin Bond 28
Caley, Frederick W. 18, 20
captains, job description 7
carnivals 36, 40, 88
Carpenter, Jack 46
Central Road 50
Chapman, Donald 71, 76, 83, 84, 85, 87
Chapman, R. 90
Chapman's builders (Hanworth) 12–13
charity football 72–3
children's entertainments 59, 61, 62

chip pan demonstration 88
church
    bells 12, 14, 19
    fires at 40, 45
    tower 45, 50, 55
Church Street flats 87
Cliff Avenue 12
Cliftonville Hotel 79
Cole, Joe 79
Collow, W. 62
Colvin-Smith, Dr 33
committees 21, 48
computers 92–3
Cook, Sidney 45
Cook, Sir Thomas A. 34, 37, 61, 63
Cooper, D. 90
Cooper, Harry 36, 37, 63
Cooper, John 84, 85, 87, 90
Cox, William 86–8
    Dutch visit 77
    incidents attended 74, 76, 83, 86–92
    pictures 74, 83, 84, 86, 90
Crane, George 33, 36, 37
Craske, Jimmy 44
Crisp, G. 87, 90
Crisp, Mr 27
Crome, Mr 36
Cromer in 1881 6
Cromer Hall estate 36, 37
Crown, R. 19
curfew 4
Curtis, J. W. 28, 30
cutting equipment 92

Dalli, Tony 72
Danaher, J. 25, 33, 36
Davey, Guy 31
Davies, Billy 83
Davies, Richard 76, 83
Davison, Daniel 25
demonstrations 17, 18, 35, 88
Dennis, Jimmy 44
Dewar, A. R. 52
discipline 7, 33
Dix, G. R. 64
Durrant, C. 33, 36
Durrant, R. 63
Dutch exchange visits 76–7

East Coast Motor Company 39, 40, 46
East House 38, 53
Edghill, Jamie
    Dutch visits 76–7
    fundraising 72–3, 79–81
    incidents attended 71, 76, 84–6, 88–90
    pictures 71, 82, 83, 84, 90

Edinburgh 5
Edinburgh House 91–2
equipment 33, 92–3; *see also* hose; breathing apparatus; ladders; reviving apparatus
Everitt, J. 33, 36, 37, 63

Fairhead, Ken 63, 65, 71, 74, 76
Farrows, Mr 65
Felbrigg Hall 35
Fire Alarm Posts 16
Fire Brigade Committee 48
fire brigades
    co-operation 24, 34, 39, 46, 47
    competitions 34, 36, 66, 70
    earliest 4
    nationalisation 33, 50–51
Fire Brigades Act (1938) 39, 43
fire engines 92
    Dennis 24
    escape vehicles 11, 38, 42, 55
    lights 39
    Merryweather steamer (1886) 6–9, 31
    Merryweather steamer (1931–36) 31
    private 10, 15
    towing arrangements 23, 28
Fire Guard 57–8
fire marks 4
Fire Services Act (1947) 59
Fire Services (Emergency Provisions) Act (1941) 51
fire stations
    1880s 6–7, 10
    1890–1905 10
    extension plans 42, 53, 54
    1905–71 10–11
    drilling yard 11
    living quarters 11, 27, 30, 53, 65
    1971 78–9
fire-watching 21
fireworks 79
Fisher, D. 90
Fisher, N. A. 63
Fisher, Ruby 45
floods 65
Folds, Les 70, 71, 76
football 72–3, 76–7
Ford, Agnes 45
Fox, A. H. 17, 18, 31
Frost, James King
    annual reports 8, 16, 18–19
    career 17, 32
    pictures 32
Frost, Phyllis 44
Fulcher, W. C. 62

Fulcher, W. J. 78
funding 8, 14–16, 24, 31

gas cylinder explosion 89–90
Gee, S. F. 41
Girling, S. 20
Golden Farm (Northrepps) 23–4
Gowing, Mr 36
Grand Hotel 19; *see also* Albany Hotel
Grange, The 12
Gray, P. 71
Great Ryburgh 47
Green, Dennis 65, 71, 76
Green, P. 87, 90
Grimble, James M. 28
Gurney, J. H. 15–16
Guy Fawkes night 79

Hammond, J. 63
Hanworth Hall 13–14
harness 20–21, 38
Harrison, E. W. 23, 25, 32, 36, 37
Harrison, George 44
Hastings, P. 25, 33, 36
Hawes, W. 25, 33, 36
Hayden, W. J. 78
Haydon, G. C. 78
Hedge, Simon 90, 94
helmets 33
Hindry, Reg 73, 77
Hoare, Samuel 15
Holt 25–7
Home Office 37, 42, 43, 44, 46, 47, 48, 51
horses 8, 13, 20–21, 28, 30
hose 26, 28, 31, 37
hospital 40
hydrants 4, 31, 40, 41, 43, 45, 48, 54

insurance companies 4, 18, 43

Jarvis, Jack 77
Jay, Gordon & Bunny 79
Jewkes, Penny 72–3
Jewson's 86–7
Johnson, Stuart & Graham 84
Johnstone, Ian 94

Kemp, C. 71
Kemp, J. J. 36, 39
Kemp, Stella 44
Kemp, Swiggy 52
Kettle, J. 17
Khalil, Harry 90
King family (Central Road) 50
Kirby, A. 25, 44
Kirby, Eric 33, 36, 44, 66
Kirby, John Bentley
 accommodation 27
 career 17, 28–9
 death 56
 pictures 25, 28, 29, 33, 36
 retirement 53
Kirby, Mrs 29
Kirk, Alfred 70, 71, 76

Knapton, George 71
Knights, A. I. 51–3

ladder drill 18–19
ladders 38
Laundry 39
Lax, Martin 74
Lee, G. 85, 87, 90
Leeds, Jane 60–61
Leggett, F. W. 17
Lewis, Mr (of Holt) 25–6
Life Saving Apparatus Co. 29
lifeboat 51, 52–3, 83
lorries 23, 29
Louch, Norman 78
Love, F. W. 17, 25
Lusher, W. J. (Billy) 25, 30, 33, 36

Mckenzie, W. D. 63
Madurodam 77
Manor Farm (Bessingham) 41
Marsh, Barry 77
Marsham, heath fire 64
Martin, Alf 44
Mason, W. 72
Meadow Playgroup 80
Meadow, The 37, 56
messengers 12, 13, 49, 58
Mill Farm (Thurgarton) 40
Mitchell, Harry 66
 appointed Auxiliary 38
 incidents attended 63, 64, 71
 pictures 36, 44, 62
 succeeds Randall 62
Mitchell, Ray 61
model village 77
Morris, Jeff 76, 84, 85, 88
Morris, S. 85, 90
Mount Street 10
Muirhead, Frank 51
Mundesley 38
Murrell, William 10

National Fire Brigades Association 38, 55
National Fire Service 50–51
nationalisation 33, 55
Newhaven Court 68–70
 tennis courts 64
Newland, Geoffrey 71, 76
Nockels, Sammy 25, 33, 36, 44
Norfolk Fire Brigades (Competition) Association 37
Norfolk Textured Yarns 84–6
North Walsham 93
Northrepps 23–4, 29
Northrepps Hall 15–16
Northway, Mr 83
Norton, Bob 66, 70, 71

open days 79; *see also* demonstrations
Overstrand 88–9
Overstrand Court Hotel 83
Overstrand Hotel 19

Palmer, George 71

Palmer, R. W. 13, 19, 20–21, 30
Palmer, W. R. 21
Parish Fire Engine Act (1898) 14
Parish Pumpers Act (1708) 4
Parry, Bill 72
pay 19
Payne, H. 17
Pearce, Harry 66, 71
Pearson, R. L. (Bob) 72, 78
Penn, Pat 77
pier 42, 43, 46, 55, 84
Pipe, Pat 77
Pooler, John 81
Pope, Joe 66, 70, 71, 76
Priest, Edward Raven 6, 28
Punchard, Peter 73

Quinn, Derek 73

railways 6, 33, 40
Randall, T. L. (Theo) 59–63
 at Frost's funeral 32
 career 51, 53, 62–3
 incidents attended 29, 63
 and inter-brigade competitions 34, 36
 pictures 25, 33, 34, 36, 37, 44, 61, 62
 retirement 62
Randell, C. 25, 33, 36, 37
Ray, Andrew 73
Read, W. G. 63
recruitment 8
Red Lion Hotel 14
regulations 6–7
retained firemen 19
reviving apparatus 37
Reynolds, Mr 38
road traffic collisions 92
Roberts, Dave 79, 84, 85, 87, 90
Roughton 35
 heath fire 40
 mill 13
Roughton Road signal box 33, 40
Rounce, J. 36, 40
Rounce and Wortley 23
Royal Links Hotel 31, 40, 63
Royal Links Pavilion 82
Royall, Alfred 33, 36, 37, 63, 66, 71
Royall, Harry 36
Runton 31
Rust's 71

St John Ambulance 63
salvage 52
Sandford, Henry 10
Sandford, W. G. 28
Seago, C. G. 63
Sennowe Park 34, 61
sentries 45, 48
Sheringham 24
Sheringham Lodge 12, 18
Sidestrand 82
sirens 43, 55, 92
Smith, Edward Girling 44
Smith, F. 33, 36
Smith, Mark 79, 84, 85, 87, 90

smoke helmets 33
snorkel 85, 91
Southrepps 30
Spink, Colin 77
Spurrell, Denham 39
statuettes 80–81
Steele, Tommy 72–3
stirrup hand pumps 49
Stonell, T. 36
Storey, W. E. 63
subscriptions 14–16
Suffield Park 16, 31, 44
 Infants School 79
Suffield Park Hotel 46, 56
Swann, C. (engineer) 25, 33, 36, 37
Swann, Cecil J. (Duckie) 25, 33, 36, 37, 40, 63, 65, 66

Tansley, H. H. 37, 49
Taylor, Pat 76
Teddington, SS 51
telephones 31, 39, 40, 92
telephonists 44, 45
thermal imaging 93
Thetford, E. 71
Thetford, J. 63
Thompson, Aubrey 44
Thurgarton 40
Tooley Street warehouse 5
Tovell, Francis George 46
Town Hall 10
training 42, 49, 93
Tribe, James 55, 58
Trimingham 31
Trollope 44, 53, 54
Tubby, Terry 77
tyres 42

uniforms 7, 22–3, 63

vehicles: Chrysler 46; *see also* fire engines

Walker, C. S. 63
Walker, Kenny 73, 77
Wallace, Don 78
Warner (assistant engineer) 12
Warren Court 89–90
water supply 14, 15–16, 31, 53, 65, 66; *see also* hydrants; pier
West, Richard 84
West Street 56
Weybourne 41, 47
whistle at laundry 39
White, Arthur 44, 63
White Horse Inn 46–7, 65
Wickers, Peter 66, 71, 76, 84
Wilkin, Mr 30
William the Conqueror 4
Willins, A. E. 33
windscreens 37
Wolterton Hall 31, 65–6
World War 1 21
World War 2 43–58
Wright, H. J. 40